Taggart and Morrell Bullock

The County of Eaton, Michigan

Topography, History, Art Folio and Directory of Freeholders

Taggart and Morrell Bullock

The County of Eaton, Michigan
Topography, History, Art Folio and Directory of Freeholders

ISBN/EAN: 9783743423930

Manufactured in Europe, USA, Canada, Australia, Japa

Cover: Foto ©berggeist007 / pixelio.de

Manufactured and distributed by brebook publishing software (www.brebook.com)

Taggart and Morrell Bullock

The County of Eaton, Michigan

THE

COUNTY OF EATON,

MICHIGAN;

Topography, History, Art Folio

AND

Directory of Freeholders.

BULLOCK, TAGGART & MORRELL,
TOPOGRAPHERS AND PUBLISHERS;
1895.

TABLE OF CONTENTS.

Topographical.

Historical and Statistical.

Printed at the Eaton County Republican Printery, Charlotte.

Map of Eaton County

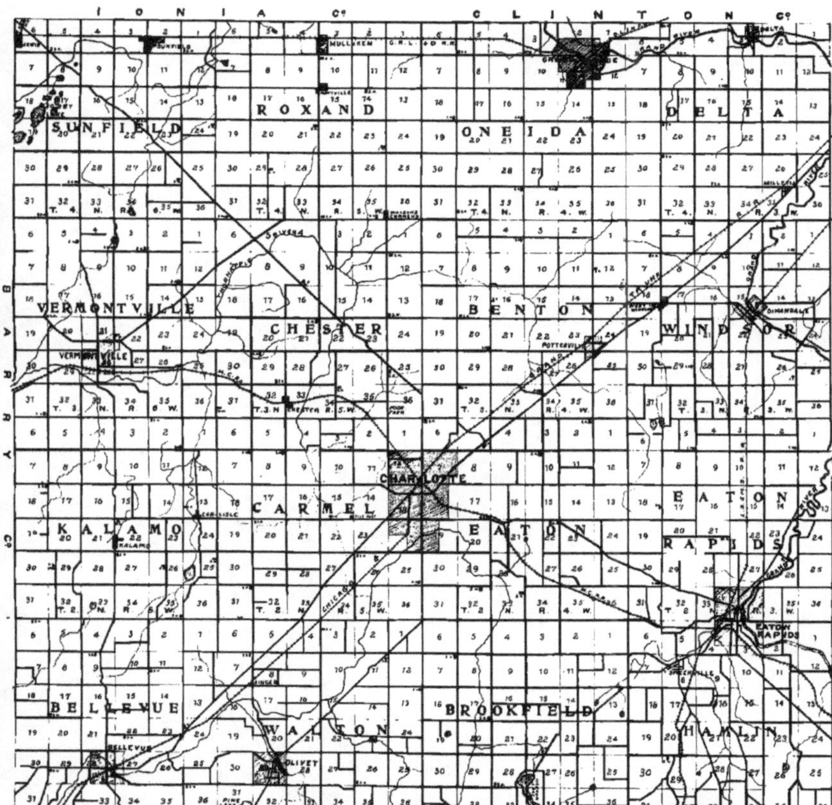

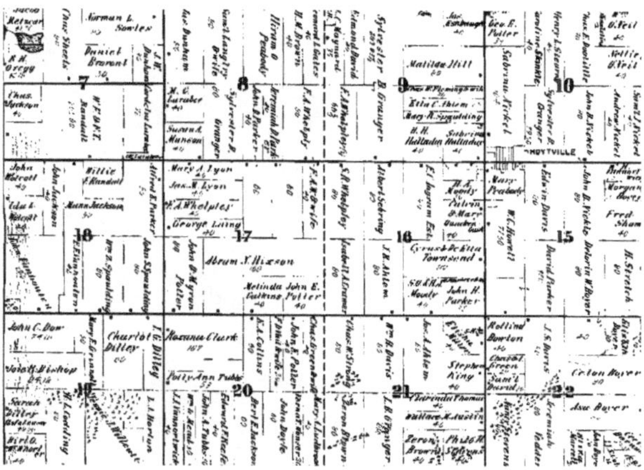
7
8
9
10
15
16
17
18
20
21
22
Norman L. Sowles
Daniel Bronont
Matilda Hill
HOYTVILLE
Sabrina Nickel
Mary A. Lyon
George Laing
Abram N. Hixson
Melinda Calkins
John E. Potter
Edwin Davis
Charlot Dilley
Rozanna Clark
Stephen King
Asa Boyer

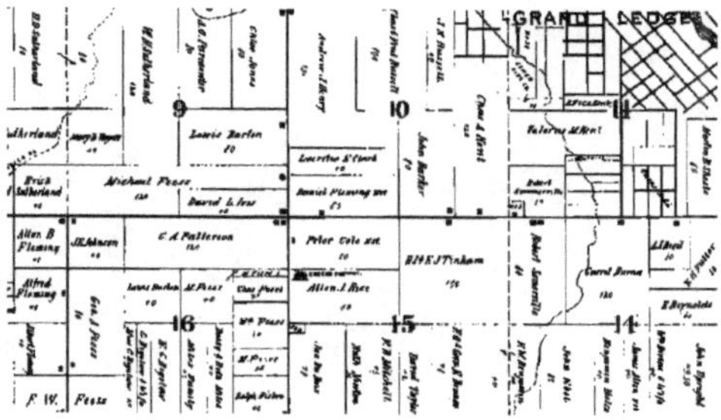
GRAND LEDGE
9
10
11
16
15
14
Lewis Barton
Michael Pease
C. A. Patterson
Peter Cole est.
Allen J. Rice
Allen B Fleming
Alfred Fleming
F. W. Pease

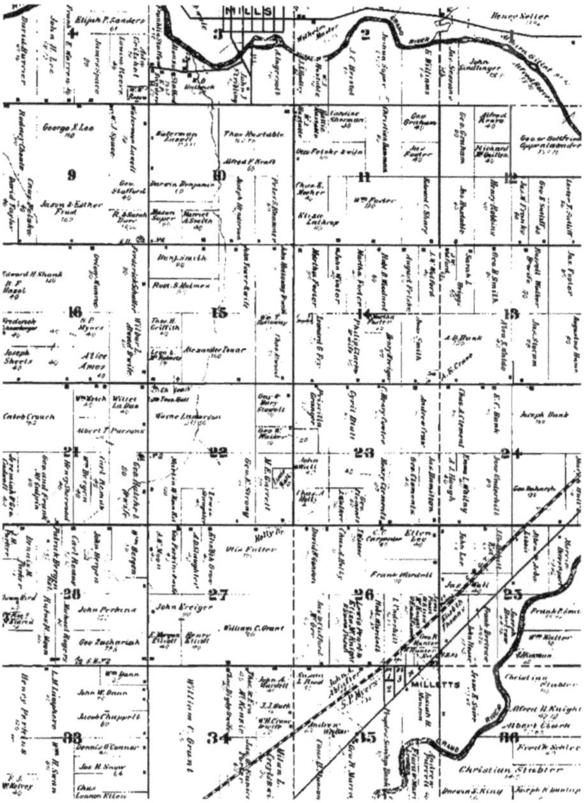
MILLS
MILLETTS
George X Lee
Caleb Crouch
Joseph Bunk
Otis Fuller
John Perkins
William C. Grant
Frank Wardell
S.P. Myers
Christian Stabler
Henry Perkins

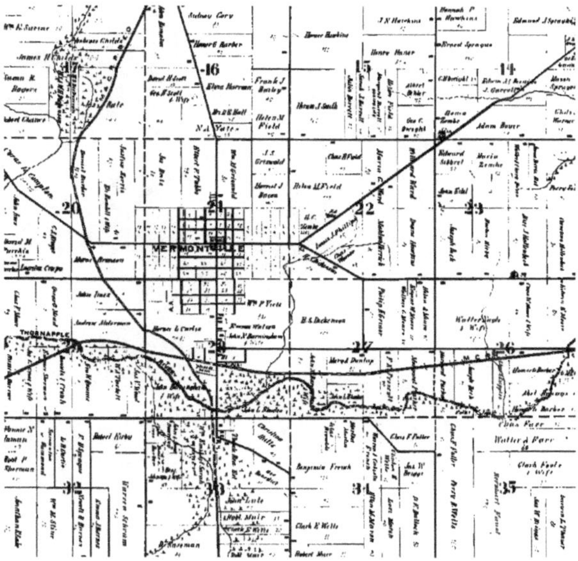

17
16
15
14
20
21
22
23
28
27
26
33
34
35
VERMONTVILLE
THORNAPPLE

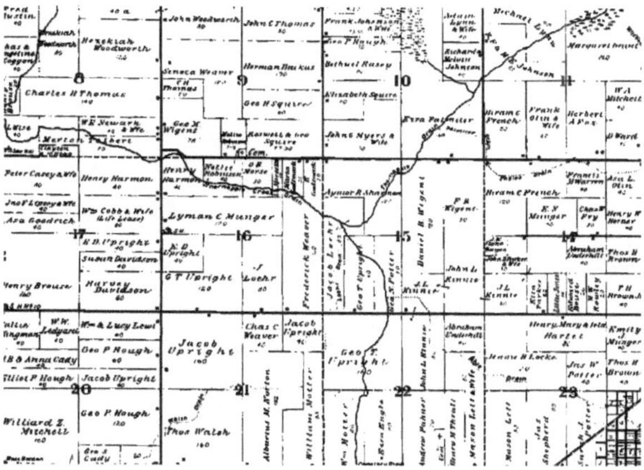

Hezekiah Woodworth
John Woodworth
John C Thomas
Frank Johnson & Wife
Geo P Hough
Seneca Weaver
Bethuel Rasey
Charles H Thomas
Geo H Squires
Elizabeth Squire
Hiram C French
Frank Olin & Wife
Herbert A Fox
W.A Mitchell
W E Newark & Wife
Geo H Wigent
John Myers & Wife
Peter Casey & Wife
Henry Harmon
Ana Goodrich
Lyman C Munger
E D Upright
Susan Davidson
Harvey Davidson
Henry Brouse
G T Upright
J Loehr
Frederick Weaver
Jacob Loehr
Geo T Upright
Daniel B Wigent
J L Kinnie
John L Kinnie
Abraham Underhill
Jacob Upright
Chas C Weaver
Wm & Lucy Lewis
Geo P Hough
Jacob Upright
Williard Z. Mitchell
Thos Walsh
John L Kinnie
Andrew Palmer
Mason Lett
Jas W Potter
Thos H Brown
8
9
10
11
17
16
15
14
20
21
22
23

Mary A Robbins
DIMONDALE
14
Edward G Beckwith
Chas Holcomb

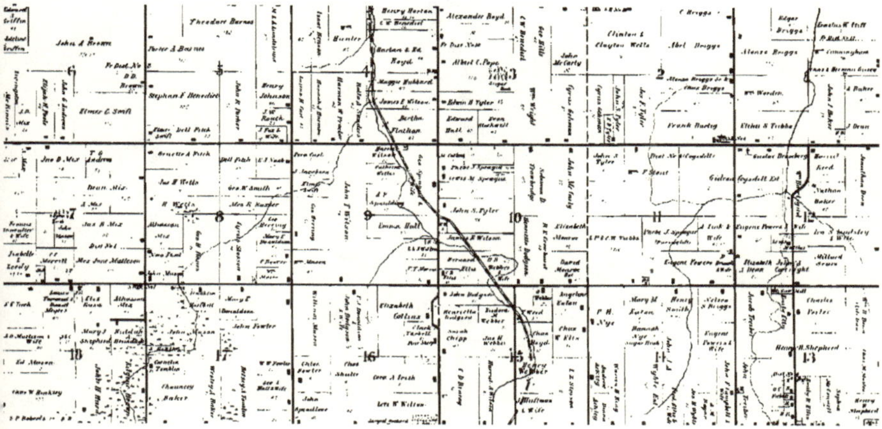

Theodore Barnes
Alexander Boyd
John S. Tyler
Emma Hall
Henry B. Shepherd

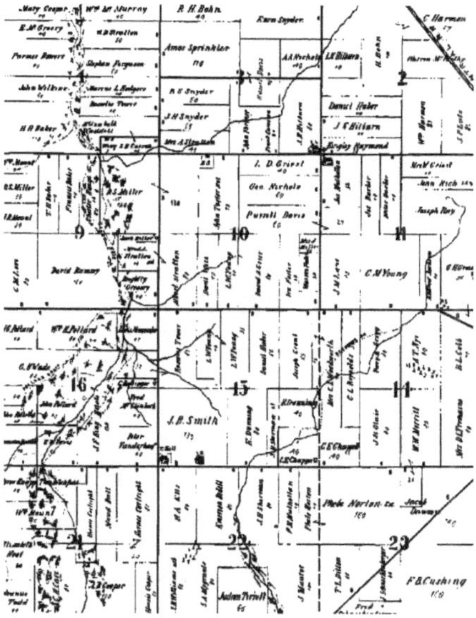
Amos Sprinkler
J. D. Griest
Geo. Nichols
C. M. Young
J. B. Smith
F. B. Cushing
2
9
10
11
14
15
16
22
23

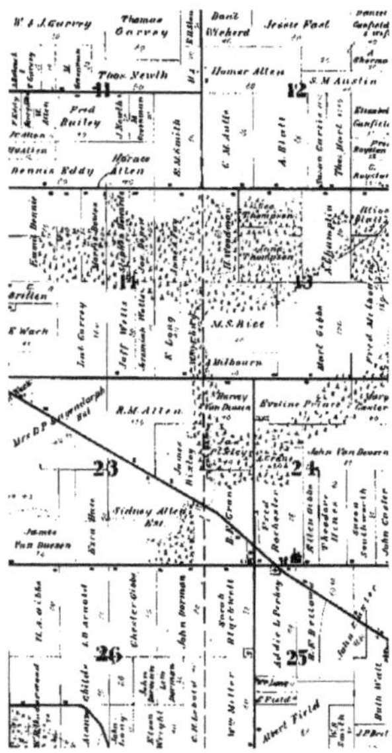

Thomas Garvey
Jessie Paul
Thos. Newth
Homer Allen
11
12
S. M. Austin
Fred Bailey
Horace Allen
Dennis Eddy
14
13
M. S. Rice
R. M. Allen
23
24
John Van Deusen
Sidney Allen Est.
James Van Deusen
26
25
Albert Field

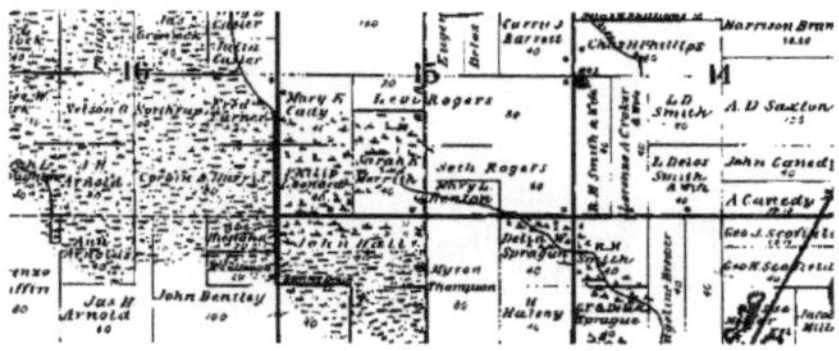
16
15
14
Currie J Barrett
Harrison Brown
A. D Saxton
L.D Smith
John Canedy
A Canedy
Mary K Cady
Levi Rogers
Seth Rogers
Nelson B Northrup
John Hall
Myron Thompson
John Bentley
Jas H Arnold
Delos Smith & Wife
Geo J. Scofield

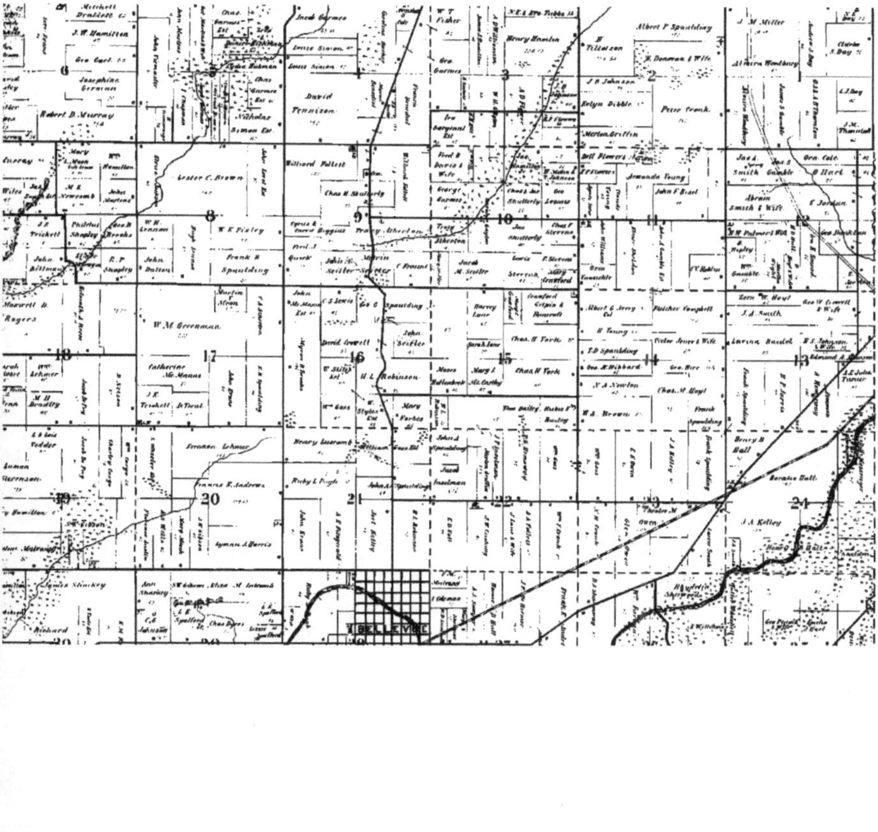

J. W. Hamilton
Robert D. Murray
David Tenniszon
Henry Hamlen
Albert P. Spaulding
Peter Frank
Lester C. Brown
W. K. Pixley
Frank B. Spaulding
W. M. Greenman
Catherine McManus
H. L. Robinson
Chas. H. York
Fletcher Campbell
J. A. Smith
T. D. Spaulding
Geo. M. Hibbard
N. A. Newton
Chas. M. Hoyt
Henry Luscomb
Francis E. Andrews
Lyman A. Harris
J. A. Kelley
Horace Hall
BELLEVUE
6
8
9
10
11
12
13
14
15
16
17
18
19
20
21
2
3
24

Stephen B Johnson
A M Nye
C E Miller
E O Nye
A B Miller
M Warner
G V Pellow
M Robinson
M L Sherman
C A Sherman
C E Scott

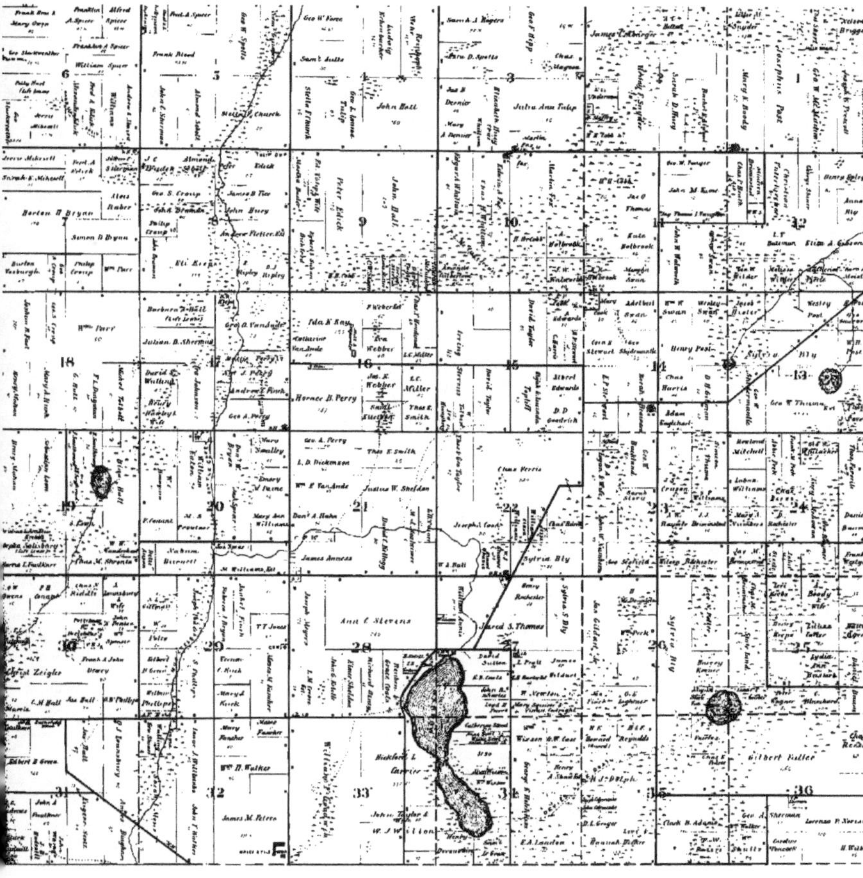
Geo W. Spotts
Sam'l Sutts
John Hall
Peter Edick
Ida K Kay
Horace B. Perry
Thos E. Smith
Justus W. Sheldon
James Anness
Ann E Stevens
Jared S. Thomas
Sylvia Bly
Henry Post
Adam Englehart
Gilbert Fuller
Clark H. Adams
James M. Peters
Wm H. Walker
Joseph Cook
Lorenzo P. Norton
Josephus Post
Mary E Boody
Burton H Bryan
Simon D Bryan
Eli Keep
Wm Parr
Geo A. Perry
L. D. Dickenson
David Taylor
Irving Stevens
Thos M. Streeter
Christ Zeigler
Asa Ball
E.A. Landon
Sylvia S Bly
Chas Harris
Geo W Thomas
1
3
4
5
6
7
8
9
10
11
12
13
14
15
16
17
18
19
20
21
22
23
24
25
26
27
28
29
30
31
32
33
34
35
36

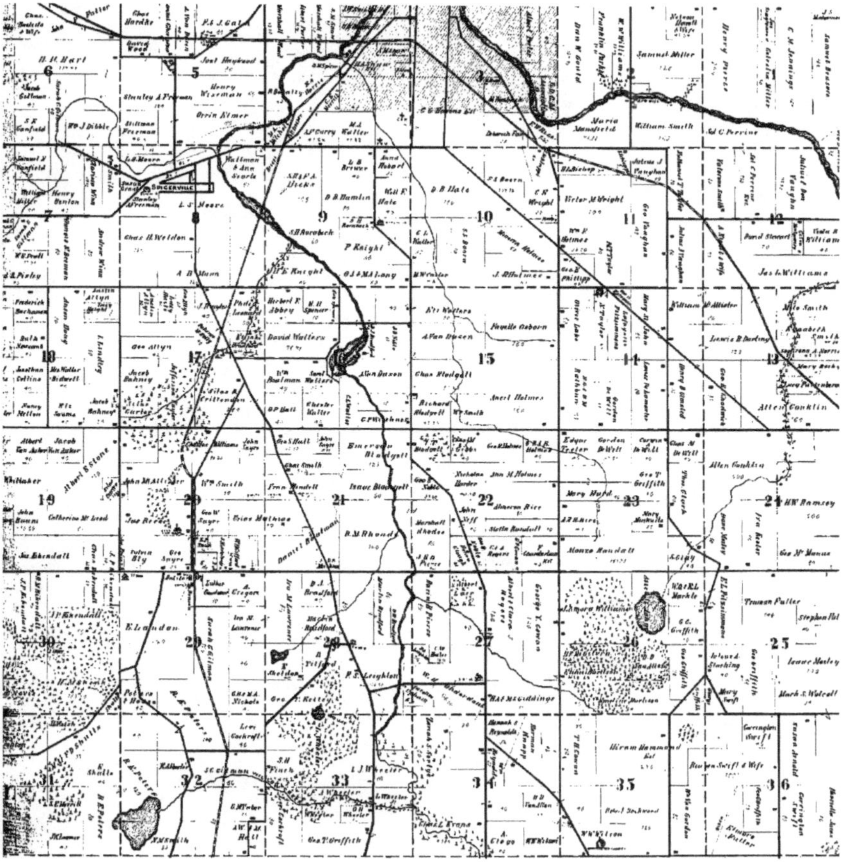

SPICERVILLE
H. H. Hart
Joel Haywood
Henry Wiseman
Orrin Elmer
Stanley A. Freeman
L. S. Moore
Chas. H. Weldon
A. B. Mann
D. B. Hamlin
D. B. Hale
P. Knight
David Walters
A. Van Dusen
Favelle Osborn
Allen Dusen
Chas. Blodgett
Ancel Holmes
Allen Conklin
Samuel Miller
William Smith
Maria Mansfield
Henry Pierce
Jas. L. Williams
Geo. Allyn
Emerson Blodgett
Isaac Blodgett
B. M. Rhoads
H. W. Ramsey
Mary Hurd
Alonzo Randall
Truman Fuller
Isaac Marley
Hiram Hammond
E. Landon
P. S. Leighton
I. J. Wheeler
Geo. T. Griffith
Reuben Swift & Wife
6
5
3
7
8
9
10
11
12
18
17
15
14
13
19
20
21
22
23
24
30
29
28
27
26
25
31
32
33
34
35
36

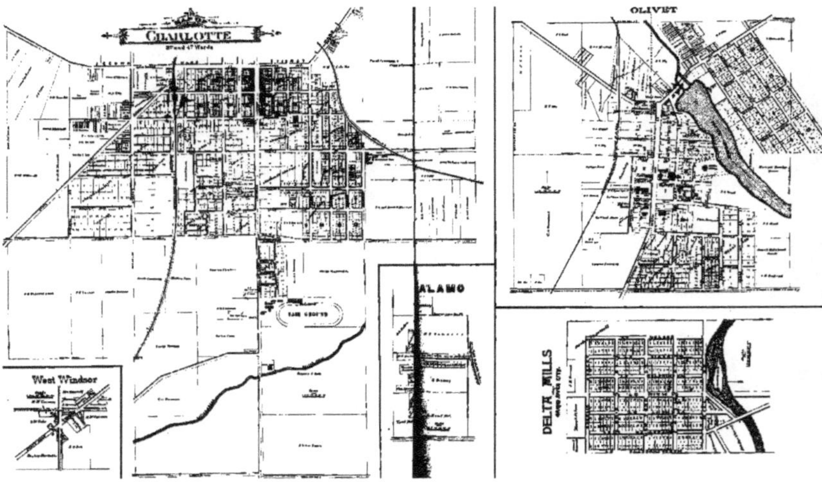
CHARLOTTE
West Windsor
ALAMO
OLIVET
DELTA MILLS

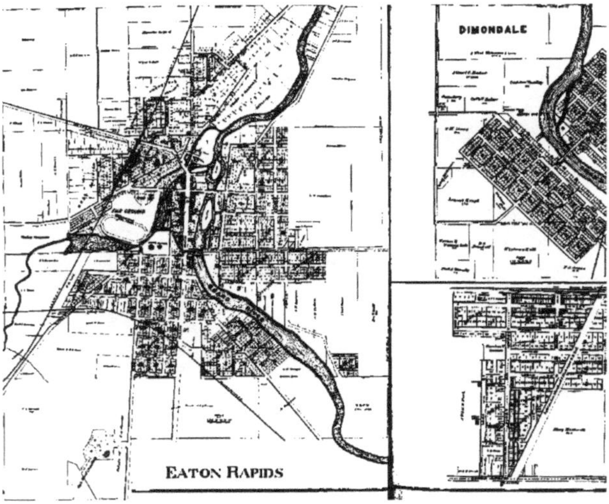
DIMONDALE
EATON RAPIDS

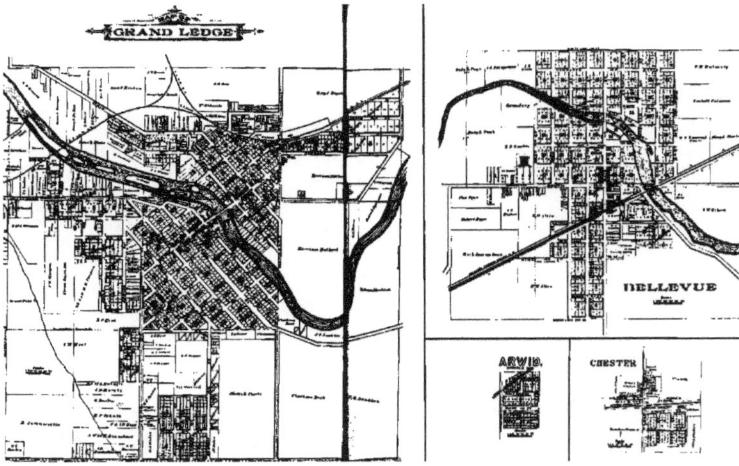
GRAND LEDGE
BELLEVUE
ARWID.
CHESTER

VERMONTVILLE

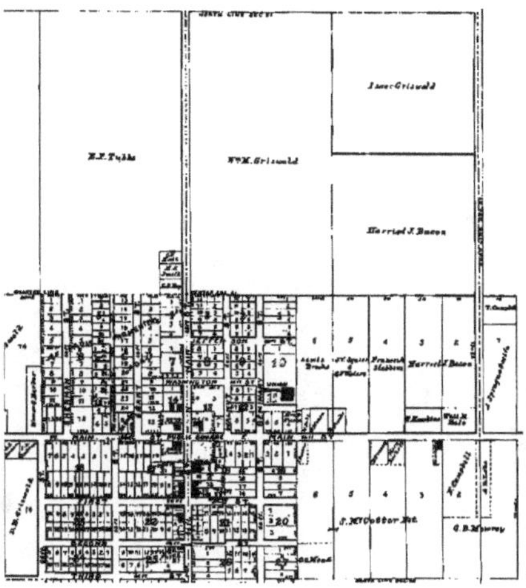

MULLIKEN

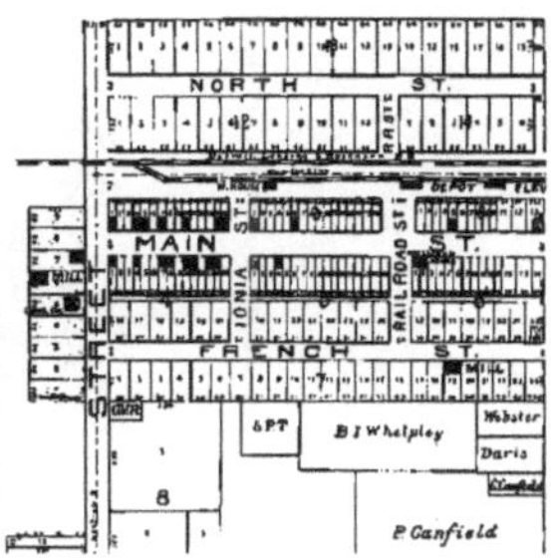

BENCH AND BAR.

1 Clement Smith, Circuit Judge
2 Jacob L. McPeek, Judge of Probate
3 Frank A. Dean, Counsel to Naples, Italy
4 J. M. C. Smith, Charlotte
5 Horace Maynard, Charlotte
6 G. C. Fox, Charlotte
7 Le Roy Jones, Charlotte
8 Edw. K. Shaw, Eaton Rapids
9 Milton A. Betts, Charlotte
10 J. B. Hendee, Eaton Rapids
11 Geo. L. Hauser, Charlotte
12 L. H. McCall, Prosecuting Attorney
13 J. C. Nichols, Charlotte
14 Geo. Huggett, Charlotte
15 J. M. Powers, Charlotte
16 L. H. Corbin, Eaton Rapids
17 B. T. Jones, Grand Ledge
18 W. R. Clarke, Grand Ledge
19 C. O. Markham, Eaton Rapids
20 Henry D. Jones, Grand Ledge
21 W. Stine, Charlotte
22 Cassius Alexander, Grand Ledge
23 W. S. Morey, Bellevue
24 R. E. Wood, Grand Ledge
25 Hon. H. A. Shaw, Eaton Rapids

COUNTY OFFICERS.

1 J. B. Smith, Treasurer
2 M. E. Newcomb, Drain Com.
3 J. L. Wagoner, School Com.
4 Frank M. Green, Sheriff
5 R. A. Carlber, Register of Deeds
6 Geo. Decke, Clerk

BOARD OF SUPERVISORS.

7 Warren Davis, Carmel
8 Stephen Benedict, Kalamo
9 Joseph Bacon, Charlotte
10 Albert Ford, Mayor, Charlotte
11 C. T. Hartson, Mayor, Eaton Rapids
13 Albert Shotwell, Windsor
14 H. L. Curtis, Vermontville
15 S. B. Evans, Bellevue
16 S. W. Mapes, Walton
17 C. L. Carr, Eaton Rapids Twp
18 J. T. Fuller, Hamlin
19 Wm. M. Beekman Charlotte
20 L. D. Dickinson, Eaton
21 J. B. Rudesill, Brookfield
22 J. W. Dunn, Delta
23 Dwight Backus, Benton
24 J. S. Hamlin, Eaton Rapids City
25 Chas. W. Dean, Chester
26 J. H. Bera, Sunfield
27 John Ewing, Oneida

EATON COUNTY, MICHIGAN.

JANUARY 11, 1805, Congress passed an act providing for the organization of Michigan Territory, and the act was made effective on June 3d of the same year. The territory was formed from a portion of Indiana, and consisted of the Lower Peninsula only, the remainder of our present State being still attached to Indiana and Illinois.

The Territorial Capital was fixed at Detroit, a small French trading village of log huts. President Jefferson appointed as officers Gen. Wm. Hull for Governor, and Hon. A. B. Woodward, for Presiding Judge. Governor Hull filled his position with honor and credit until August 16, 1812, when he surrendered fourteen hundred troops and the whole of Michigan Territory to a few hundred British troops. For this act he was stripped of all official title and Gen. W. H. Harrison was appointed as his successor.

Gen. Harrison exercised gubernatorial authority over the Territory of Michigan until October 13, 1813, when he resigned in favor of Col. Lewis Cass. By various appointments Col. Cass retained this position until he was called to a seat in President Jackson's cabinet in 1831. His career as Governor of Michigan, noted as the longest, the most peaceful, the most effective in developing out of a wilderness a beautiful and prosperous state, was ended, but his memory is still fresh in the minds and hearts of many very old citizens. The names of streets, townships, cities and counties testify that his place in our history will not be forgotten. A really useful, heroic man lives forever.

In 1815 Congress established a base line and principal meridian from which Congressional Townships and Ranges might be surveyed and numbered. The next year public lands were surveyed in the vicinity of Detroit and were offered for sale soon afterwards at the Detroit Land Office. From that time on the State has been gradually surveyed and opened for the people until at present only a small area of the Upper Peninsula is known as public land.

About this time the formation of counties began. At first the County of Wayne included about the whole of the Territory of Michigan, but from time to time portions were cut off and called by new names.

On the 29th day of October, 1829, the Legislative Council of Michigan passed an act forming the County of Eaton, and at the same time twelve other counties, comprising a large scope of country, the richest and most populous of Southern Michigan today. This was the first year of the administration of Andrew Jackson. John H. Eaton was Secretary of War in his cabinet and from him the County received its name. The Counties of Berrien, Barry, Ingham and Branch were also named for members of his cabinet; Jackson, in honor of the President himself and Cass and Calhoun for distinguished Democratic statesmen of the day.

But while these formations took place in the year 1829, their several populations were very small, and for this and other reasons the organization of these Counties did not occur until years afterwards and at various times. Eaton County was not organized until December 29th, 1837.

The Ordinance of 1787, establishing the Northwest Territory, provided that when a Territory contained a population of 60,000 it should, upon application, be admitted as a member of the Union. In Michigan the preliminary steps were taken in 1834. A census was taken showing a free white population of 87,278. The Legislative Council passed an act authorizing a convention to be held in Detroit for the purpose of framing a state constitution.

This convention, of 89 delegates, convened at the specified time and framed a constitution, which was submitted to the people and adopted by a vote of 6,299 to 1,359. At the same time a full set of state officers and a Legislature were elected to act under the new constitution. Hon. Steven T. Mason was chosen first Governor of "The State of Michigan."

But before admission could be granted, a certain boundary trouble, called the Toledo war, had to be adjusted. The people of Michigan claimed that the ordinance of 1787 described their boundary as extending South of the Maumee river, and giving them the port of Toledo. Ohio claimed that this was not the intention of Congress, and that this description had been based upon inaccurate maps. Congress compromised the matter by giving to Ohio the disputed territory and to Michigan, as its equivalent, that which is commonly known as the Upper Peninsula. The bill of admission was granted the State January 26, 1837.

The state constitution did not fix upon any permanent location for the State Capital, and in 1847 the Legislature, after a long discussion, decided upon Lansing, a small village in Ingham County, as the proper place for its location.

But while these state affairs were transpiring pioneer farmers, merchants, landlords, mill men and manufacturers were rapidly putting in their appearance at Bellevue, the gateway into the deep forests of Eaton County.

In making the journey to his new home the immigrant found travel comparatively easy through the oak openings from Kalamazoo, Battle Creek and Marshall. But at Bellevue was the jumping-off place, into the wilderness, and the coming-out place of the burrowing settler, where he was once more in the light of day. In speaking of the early inhabitants we refer, of course, to the whites, for, previous to 1840, Pottawattomie and Ojibewa Indians were here in large numbers. They were the original owners of the oak, walnut, sugar and maple forests that grew in this favored portion of the State. If they had remained here to the exclusion of the whites the great development of the County would not have occurred, for the civilized Indian is an Indian still, with little capacity for the development of a country.

The land occupied by these Indians was fine territory for hunting and fishing, and the other pursuits peculiar to their tastes, but they were early compelled to surrender to the more capable whites. The Pottawattomies were removed by the government in 1840, to territory beyond the Mississippi. Gen. Brady, who was in charge of the work, sent his agents and soldiers through the woods in search of them. Pursued by the troops, and as unwilling to leave their forest homes as we would be to leave ours, they met in council just west of Bellevue; mounted on their ponies, which stood arranged in the form of a circle, a solemn consultation was held. When they separated, one company of them fled north into the forest, but was overtaken by the government troops, brought back to Marshall, and eventually banished to land reserved for Indians in the far west.

The last night of their stay was indeed a night of sorrow. The squaws moaned; and the men wrapping themselves up in their blankets, bore in silence their grief; it was hard, even for an Indian, with his stoical nature, to endure. The government had wisely refused them the poor consolation of whisky.

EARLY SETTLEMENTS.

The first land entry in Eaton County, according to the Tract Book, was made in 1829, by A. Sumner, on Section 30, of Vermontville. No other entries were made that year nor the next. In 1831 H. Mason made an entry on Section 2 of Oneida; in 1832 three entries were recorded and a like number in 1833.

Reuben Fitzgerald, a most familiar name in the early history of the County, was the first actual settler. He arrived in July, 1833, and located in the Township of Bellevue, as it was afterward called, when divisions were made and names given them. This useful pioneer was born in Montgomery County, Maryland, February 23, 1800. He began life for himself as a farm hand; then for a year or two operated a farm on shares; later, he entered a foundry, and with the money he was able to save during his seven years experience as a moulder, he bought a small farm. He sold this, emmigrated to Michigan, built himself a bark shanty and soon became one of the best known citizens of the new country. It is impossible to write a history, however brief, of Eaton County without a reference to his honorable career. He died July 20, 1878.

Very soon after the settlement of Mr. Fitzgerald others came and Bellevue, as before mentioned, receiving the greater part of the early influx of immigration, became in a very short time the metropolis of Eaton County.

The birth of the first white child occurred here when Sarah, daughter of Capt. Fitzgerald, was born, November 12, 1834. Here a Mr. Baker met his death, and was the first white man buried in the county. The first town meeting was held here in the spring of 1835. The voters of the county, four in number, assembled for the first time in a log shanty, called in New England fashion, the meeting house. This building was church, school house and town hall combined, and the small but historically interesting company of original voters consisted of Capt. Fitzgerald, S. Hunsicker, Calvin Phelps and John T. Hoyt, the latter of whom was chosen clerk of the election.

The officers of the election took their seats; Calvin Phelps stepped to the front of the cabin, took off his hat and in a loud voice proclaimed, "The polls of this election are now open." and warned all men under penalty of the law to keep the peace; the humor of which was fully appreciated by those present. There were more offices than voters to receive them; so the minor ones were given to the ineligibles who had gathered to attend the town meeting from five hundred and seventy-six square miles of territory. The election board waited until the legal hour for closing the polls before ascertaining the result of the election.

Lawrence Campbell, in 1836, built and kept the first hotel, called the Bellevue Village Inn. The oldest post office in the county is situated here, and John T. Hoyt was made its first postmaster. His commission was dated May 2, 1835, but it was near the close of the following summer before it reached him. It cost twenty-five cents to send a letter when Mr. Hoyt was postmaster. People did not always pay in advance for the carrying of letters and Mr. Hoyt says, "while I had the office I lost twenty-five dollars by trusting postage.

The first great public enterprise was the construction of a bridge across Battle creek in the Village of Bellevue, and the laying out of the Ionia and Bellevue road in 1835. A general subscription was taken by J. T. Hoyt. He called upon the more prominent citizens first, receiving fifteen dollars from Reuben Fitzgerald, six dollars from Daniel Mason and twenty dollars from Sylvanus Hunsicker. In all one hundred and fifty-five dollars were raised, and the road was opened to Thornapple river. It was afterwards opened to Ionia by the citizens of Vermontville. Bellevue gave a Fourth of July celebration in 1835. Rev. Asa Phelps, standing on Reuben Fitzgerald's wagon house, read the Declaration of Independence to the citizens, who then marched in a procession to the home of J. T. Hoyt, where they partook of such viands as each family of the village had brought for the occasion. The first settlers were from New York, Massachusetts and Vermont. They were wide-awake, stirring and shrewd in a deal, and soon began to look after commercial growth, as well as the more primary business of farming. Among the earliest were J. Pond, Caleb Woodberry, grocers; D. F. and J. F. Hinnman, general dealers; Abner and William R. Carpenter, grocers; Major Elija Bond, grocer and dealer in general merchandise. Mr. Woodberry started a tannery, David Lucas a painting and chair manufacturing business, I. E. and J. B. Crary built a flouring mill with two runs of stone. Hiram Ovenshier built the pioneer saw mill of Eaton County. Farms were rapidly cleared and improved and roads built. Before the change to a permanent seat of justice had been made the county business was transacted at Bellevue. It was the headquarters of good society, the center of learning, of religious instruction, of manufacturing, agriculture and commerce. When the courts and juries gave a dignity and importance years ago Bellevue stood highest in the thought of the people. But other communities were destined to grow, and Charlotte was one of them.

CHARLOTTE.

Stories conflict concerning the discovery of the plains now occupied by Charlotte, but the fact that George W. Barnes located land here in 1832 precedes the more romantic reports of discovery by George Torry and Hannibal G. Rice. There is no doubt that these gentlemen were here in the early thirties, but they did not locate land until 1838. In 1835 Mr. Barnes sold his right and title to the Northeast Quarter of Section 18 in Town 2 North, Range 4 West, and also the East Half of the Northeast Quarter of Section 13, Town 2 North, Range 5 West to Edmund B. Bostwick, of New York City. On this land the original Village of Charlotte was platted.

Jonathan Searls and his brother Samuel were the first settlers. They found their way here from Bellevue in October, 1835, located a little Southeast of the Barnes land and built a log house, the only one within eight miles, and with their families, endured all the hardships of those pioneer days. On February 1st, 1837, Jappet Fisher arrived, and about the same time Steven Kinne and wife. The death of Mrs. Samuel Searls, in June of this year, left Mrs. Kinne the only white woman for miles around. The home of the Searls brothers soon became the headquarters of the county, and all sorts of public meetings were held there. Settlers came in rapidly during the next three years, among whom may be mentioned Simon Harding, Allen Searls, Hiram Shepherd, and Eleazer Stearns the first settler to locate within the limits of the original plat of Charlotte. Enterprises were not slow to start. Mr. Bostwick, the owner of the site of Charlotte, caused it to be platted into a village, and named the place in honor of his young wife, Charlotte; the streets and avenues were named for Francis S. Cochrane, Thomas Lawrence, Townsend, Harris, and Bostwick avenues are for the owner himself. These men were interested with Bostwick in building up the village.

The Eagle Hotel, a large block building of early renown, was constructed in 1838, on the site of the Phenix House. Its name was afterwards changed to the Charlotte House. It was burned May 20, 1862. In the summer of 1838 Jonathan Searls was appointed postmaster of Charlotte; a boy named Isaac Hill carried the mail bag through from Marshall once a week. The first school of the village was held in a small house built by a young man named LeCont, and in it was also instituted a pioneer debating society. But the growth of Charlotte was very slow in these days, owing to the want of money. The first year or two in a heavily timbered country, with all the money invested in land, with nothing but an ox team and an ax to do with, no crops to turn into money and most of the people sick with the ague, made close, cramping times, indeed. A prominent writer says that five or ten dollars in one man's pocket produced a sensation in those days. Everybody knew of it and the man was highly respected. All kinds of schemes were laid to borrow it, to sell him a watch or a rifle, or to work up some kind of trade which would bring in a little boot money, just enough to sweeten it.

Although houses were far apart, neighbors lived very near in those days. In trouble, in sickness, at weddings or funerals, everybody was there to do all that could be done; to feel all the sorrow or joy or sympathy that could be felt by those who knew and undersood each other so well.

At a meeting of the Board of County Commissioners, held March 19, 1840, it was resolved that all future business pertaining to the county or its affairs should be held at the house of William Stoddard in the Village of Charlotte. This building, which was intended for a tavern, served for five years as court house, jail, hotel and dwelling. In 1842 the office of County Commissioner was abolished and the affairs of the county transferred to the Board of Supervisors. The unique ideas of public economy as expressed by this honorable were probably due to the excessive stringency of the times. At their session in 1845 it was agreed to build a court house on the public square of Charlotte. Much dissatisfaction was expressed by the Eaton Bugle, then published in Charlotte, because said building cost nearly a thousand dollars.

In 1847 a jail of hewn timbers was erected at a cost of several hundred dollars. It stood on the site of the present jail.

As a summary of Charlotte in 1845, we quote from the Eaton Bugle: - "Improvements are now the order of the day. From our window we can number at this moment nine new buildings going up, and we hear of several others that are delayed on account of the want of materials. A new court house is going up on the public square under the steady guidance of Major Scout, and will be ready for the next term of the Circuit Court in September. Dr. Jos. P. Hall is erecting a commodious two-story dwelling on Cochrane avenue. The Messrs. Hayden are putting up a large tin, copper and sheet iron manufactory, and are preparing to go into business as extensively as any other establishment in the state. We were highly gratified to see these enterprising young men start out a traveling wagon yesterday; it is the best evidence of our prosperity. We are informed that it is the present intention of one of the proprietors to sink a tannery here this summer. A large ashery has already been erected by our friend S. E. Willett. The Messrs. Welsheimer are making arrangements to commence a saddle and harness business. Their stock and tools are already here, and in a few weeks they will be in the full tide of successful experiment. But why need we particularize? Our motto is onward! and who shall set bounds to our efforts? Commendation in behalf of Charlotte is superfluous, for to see is to love it. We confidently believe, from present appearences, that no other town in the state has fairer prospects ahead; and we know that no other can furnish so many natural beauties to feast the eye and regale the senses. Such is Charlotte, the County Seat of Eaton.

After eighteen additional years of substantial improvement Charlotte was incorporated as a village, January 7, 1863. A flaw in the description of the boundaries necessitated further action, and the Board of Supervisors at their session in October, 1863, issued an order incorporating the following territory, to-wit: The Southwest ½ of the Southwest ½ of Section 7, and the Northwest ¼, the North ½ of the Southwest ¼ and Southwest ¼ of Northwest ¼ of Section 18, Town 2 North, Range 4 West; also Southeast ¼ of Southeast ¼ of Section 12, and Northeast ¼, the North ½ of the Southeast ¼, the Southeast ¼ of the Southeast 1-4, and the East 1-4 of the Northwest 1-4 of Section 13, Town 2 North, Range 5 West.

It became a city March 29, 1871, with the Hon. E. S. Lacey as its first Mayor. At present the population of Charlotte is 4,300. It boasts of two National banks, six school building, ten churches, two railroads, a half dozen good prosperous manufacturing establishments, three newspapers and a hundred other business institutions. Its citizens enjoy the privileges of electric and gas lights, of water works, of beautifully paved streets, which are shaded on either side by overhanging maples, so that it might now with propriety be called the "Forest City," instead of the "Prarie City." Its taxable property, as shown by the returns of 1894, amounted to $2,225,000.

On August 17th, 1837, James Gallery came to Eaton Rapids. There were then only three dwelling houses in the place. The streams were not bridged, the timber was uncut, the ground uncultivated, the Indians rowed up and down the river in their canoes, and trails instead of highways led off into the forest. Amos Spicer, Benjamin Knight and C. C. Darling, with their families, were the only persons here at this time. But a dam had already been built across Spring Brook and the frame of the grist mill which is still standing had been partly enclosed. An event of considerable importance was the establishment of a post office, in 1837 or 1838, with Benjamin Knight as the postmaster. When Mr. Gallery arrived the nearest grist mills were at Jackson, but about January 1st, 1838, the mill at Eaton Rapids was started. On the corner where the Anderson House now stands Benjamin Knight erected the first store building. His little store developed into a large and successful business enterprise, In 1884 Mr. C. C. Darling had a small grocery in a shanty. He sold whisky, but was very cautious about disposing of it to the Indians who were encamped close by. In the summer of 1842 the dam across Grand River was built and the race that connects the river and Spring Brook was dug. The mill received the addition of two runs of Burr stones and a set of merchant bolts. Some seven years later Mr. Sterling located in Eaton Rapids, and in company with Mr. Seelye, entered the mercantile business.

The Frost House was originally occupied by the dry goods firm of Frost & Daniels. The south part of the building was erected in 1852 and the north part at a still earlier date. In 1870 Mr. Frost opened the entire building as a hotel and bathing establishment. Dr. Morris Hale became its proprietor in 1875. The Vaughan House was opened for business in 1872 by Pantlind & Pickering. It was a fine three-story brick with a basement under the whole, and contained 125 rooms. In 1871 it was totally destroyed by fire and has not been rebuilt. The Anderson House, an elegant, four story and basement brick hotel, was built in 1874 by W. H. Dodge. Col. G. M. Anderson, after whom the house was named, raised by subscription three thousand dollars towards its construction. This house is well furnished and fitted with an elevator. Connected with it is the Arcana well, 192 feet deep. It has a greater flow of water than most of the other wells in the place. There are a number of these mineral wells in Eaton Rapids, and their value in the treatment of certain diseases, has been clearly established. The first mineral well was sunk in 1869, and the discovery of its character created considerable excitement. The original plat of the village of Eaton Rapids was laid out July 19, 1838 by Amos Spicer, Pierpont Spicer, Christopher Darling and Samuel Hamlin. It became a village by act of the legislature April 15, 1871. In 1881 it received a city charter, and H. H. Hamilton was its Mayor. The total wealth of Eaton Rapids is about $1,000,000.

The first permanent settler in Grand Ledge was Peter Lawson, the date of his arrival being October 18, 1848. A mile and a half west of him lived John W. Russell, about the same distance north, David Taylor, and about two miles south, Peter Bozier. These were his nearest neighbors at that time. In 1840 Abram Smith came to Michigan and eight or nine years later he and John W. Russell were granted the privilege by the legislature of building a dam across Grand River. When this was completed a sawmill was erected and put in operation. Reuben Wood visited the place in the fall of 1849 and and purchased six and a half acres of land on the north side of the river. The next spring a building was erected for Wood & Allen by Smith, Russell and Taylor, and in June a general stock of goods was opened in it; this was the first store established in Grand Ledge. Wm. Russell opened the second, and the Daniels the third. The former also built a hotel opposite the store of Wood & Allen, which ranked before its destruction by fire in 1876, as the oldest in the place. There was no postoffice at Grand Ledge until 1850, when Henry French was appointed postmaster, but it was some time later before a mail route was established. The postmaster at Lan-

Res. of Mrs. T. W. Daniels, Eaton Rapids.

Res. of Jacob Upright, Benton Twp.

Late Res. of Enoch Walter, Hamlin Twp.

Res. of Homer G. Barber, Vermontville.

Reynolds Bros. Dry Goods Store, Charlotte.

Interior View of Res. of Mrs. Emma J. Church, Florist, Charlotte.

Merchant's National Bank, Charlotte.
R. T. Church, Pres. H. K. Jennings, Cashier.

First National Bank, Eaton Rapids.

Old Camp Ground, Eaton Rapids

Main Street, Eaton Rapids

1 First National Bank, Charlotte.
2 W. P. Lacey, Cashier.
3 F. S. Belcher, Pres.
4 A. J. Ives, Vice Pres.

Scene on Grand River

Sunfield Mfg. Co.
W. G. Ramsay, Sunfield, Mich.

Walter Block, Eaton Rapids

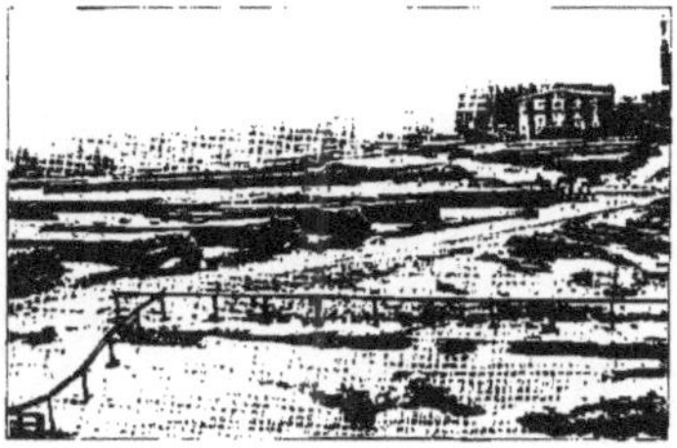

Grand Ledge Sewer Pipe Co., Grand Ledge.

ving delivered the letters to any person who happened to bring the mail bag from Grand Ledge. In 1853 a wooden bridge was built across the river. It was replaced in 1870 by one of wood and iron at a cost of $9,800. A foundry, a steam saw mill, several planing mills, and a furniture factory were added at various periods, the foundry ranking first in point of time. The Detroit, Lansing & Northern railway has included Grand Ledge since 1869 in its list of stopping places; and several mineral wells and a summer resort have attracted to the town a large number of visitors each year. The original town of Grand Ledge was laid out October 28th, 1850, and incorporated as a village by an act of the Legislature, approved April 8th, 1871. Various additions have since been made, and a city charter was granted in 1893.

In May, 1850, Isaac M. Dimond commenced to improve the water power of Grand River, on the present site of Dimondale, but in 1852 a part of the dam he had built was carried away by a freshet and the sawmill damaged to such an extent that considerable repairing was made necessary. In 1856 Mr. Dimond erected a grist mill. The village, which was laid out the same year, was named Dimondale in honor of Mr. Dimond. A. C. Bruen became his successor; the dam was repaired, the mill, which had partly fallen down, was righted and the property transferred soon after to E. W. Hunt, who continued to do an extensive custom and a fair merchant business. A postoffice, called East Windsor, was early established in the eastern part of the township. Dimondale's first postmaster was Edward W. Hunt, who was commissioned January 6th, 1876.

Nearly all the early settlers of Vermontville came from Vermont, hence the name. In 1835 the Rev Sylvester Cochrane visited Michigan and conceived a plan of colonization, which, on the 27th of March, 1836, was put into active operation. At a meeting held at that time rules and regulations for the government of the Union Colony, as it was called, were adopted. Education, religion and temperance were prominently mentioned, and it may be that the recent triumph of prohibition in Eaton County was due remotely to these pious New Englanders and the resolutions by which they were influenced. At a regular meeting of the society, held in Vermont, March 28th, 1836, it was voted that each member of the society should advance $210.50, which would entitle him to a farm lot of 160 acres and a village lot of 10 acres, or land in proportion to the amount of money contributed. This money, or its equivalent in notes, was to be furnished the agents before their departure for Michigan. William G. Henry and S. S. Church, the agents, left Vermont April 21st, 1836, to select and purchase land for the colony. It was a long and tedious journey much of; the route lay through the wilderness, and had to be traveled by stage.

W. J. Squier surveyed the site that was finally chosen for the village, and those present selected their village lots. W. J. Squier, W. S. Fairfield, Samuel and Charles Sheldon, Levi Merrill, Charles T. Moffatt, with others remained and commenced chopping and clearing, but S. S. Church returned to Vermont to get his family. The first frame house was built by W. J. Squier, and he continued to live in it until his death, which occurred in 1869. This old house has since been replaced by a fine brick structure. It was R. W. Griswold, however, who erected the first brick house in Vermontville, bringing the masons who did the work from Battle Creek. Edward W. Baker, Willard Davis, George Browning, George Squier, Martin L. Squier, Daniel Barber, Rev. William U. Benedict, Simeon McCotter and Frank P. Davis are familiar names in the history of that interesting locality, and many incidents of historical value might be given in connection with each one of them.

Wells R. Martin was the first hotelkeeper, and in company with Decatur Scoville, was also the first to open a stock of goods for sale. S. S. church brought the first mail, and a postoffice was established at the same time, with Dr. Dewey Robinson as postmaster.

Olivet, "The Athens of Eaton County," is situated in the township of Walton, on one of the most picturesque locations in Michigan. Its fitness for a village site was never questioned by man, as there are evidences of its habitation at one time by Mound Builders. The first white man in Walton township found, upon his arrival, an Indian village of about one hundred population, on the site of the present college grounds.

The village was built for the accommodation of Olivet College; hence a history of the Village of Olivet is a history of the college. Rev. John J. Shipherd, one of the managers of Oberlin (Ohio) College, conceived the idea of going again into the wilderness and building up an institution of learning. He came to Eaton County in 1843, to look after certain property situated in the Grand River Valley belonging to Oberlin College. He was delighted with the scene presented by the elevated land, the beautiful stream, the wild oak forest of Section 29, Walton township, and decided upon it as a suitable place for the location of the new school. He returned to Ohio and organized a colony of thirty-eight persons, including fourteen children and youths, and on Saturday, February 24th, 1844, the entire party arrived on the designated spot.

The first twelve months of their stay was a period of great suffering. Many were sick with the swamp fiend, ague, and returned to Ohio, but the leaders of the little band possessed the spirit to cope with the difficulties that nature presented. L. A. Green, one of the students, erected a small cottage for a study and private dormatory. This served for a chapel, recitation room and village post office. The corps of instructors included Rev. Reuben Hatch and Oramel Hosford. In 1848 a charter was granted the school under the name of the "Olivet Institute." School and village prospered under this charter and students came from many parts of the state.

In 1859 the institute was changed to a college, and from that time on no one has predicted anything but success. The village grew apace with the school and to-day exerts a great influence for good over all of Eaton County. Olivet's usual enrollment is about 300 students. Her buildings surpass those of any other institution of a like character in the state. Her instructors are liberal, broad-minded men and her alumni are, as a class, successful business and professional men and women.

Bellevue occupied originally all the territory in Eaton County. By an act approved March 11th, 1837, the township of Bellevue was divided, and the new townships of Vermontville and Eaton were set off and organized, the former including the Northwest quarter and the latter the Southeast quarter. March 6th, 1838, the Northeast half of the remaining portions of Bellevue, or the Northeast quarter of Eaton County, was set off and organized into a separate township, known by the name of Oneida. On the 15th of the same month Bellevue was further reduced by the formation of Kalamo, to include the territory in Town 2 North, of Range 5 and 6 West. No more divisions were made until March 21st, 1839, when the East half of Kalamo was set off and organized as Carmel. The East half of Bellevue was set off and organized as Walton, and the East half of Vermontville was organized into a separate township called Chester. Brookfield was formed March 20th, 1841, from a portion of the old Township of Eaton, and included Town 1 North, Range 4 West. On March 21st, 1841, Eaton was further reduced in size by the formation of Tyler, including Town 1 North, Range 3 West. February 16th, 1842, witnessed several changes. Sunfield was set off from Vermontville and made to include Town 4 North, Range 6 West. Windsor and Delta were formed from the East half of Oneida, and Eaton Rapids township was created from that portion of Eaton included in Town 2 North, Range 3 West. On the 9th of March, 1843, the township of Chester was divided and its North half was set off and organized into a seperate township, known by the name of Roxand. On the same date Oneida was cut in twain, and its South half formed into a seperate township called Tom Benton. This name was not satisfactory to the inhabitants of the town and the "Tom" was dropped by an act approved March 19th, 1845. March 14th, 1845, the Township of Tyler was united to its next Northern sister and the name of the latter—Eaton Rapids was applied to both as a whole. For nineteen years this arrangement continued, but finally, on the 25th of March, 1863, the old township of Tylor was again set off from Eaton Rapids, and organized under the name of Hamlin, in honor of one of its pioneers. No change has since been made.

There are twenty-six post offices in the county. The list, omitting those already named, is as follows: Aingur, Bismark., Brookfield, Carlisle, Charlesworth, Chester, Dellwood, Delta, Gresham, Hoytville, Kalamo, Kingsland, Millitts, Mulliken, Potterville, Roxana, Sunfield and Woodbury.

Newspapers of Eaton County.

We believe in the newspaper—daily, weekly and monthly—because it brings together the people who have something to reveal, and the people who should know what it is. Supplied with instruments that make the eye far seeing, the hand mighty, the feet swift, the ear exceedingly sensitive, the truth seeker is able to make many important discoveries; and the pure love of truth that seems so deep down, so far away, so impossible to get at, is the unfailing inspiration. Pain smitten bodies, darkened intellects, souls hidden in midnight bid him make haste, and the silent investigation goes on night and day, while the alert newspaper tells to a host of sympathising listeners of success or failure. What a crime it would be to hide the knowledge upon which life, health, happiness, and all depend! But it is not hidden. The means of revelation are now so numerous, varied and cheap, that ignorance is without excuse. Every orator has become a thousand. Talmage preaches to a few hundred on Sunday, and to millions on Monday morning. Surely the opportunity to know implies responsibility: the man who goes up from the nineteenth century, to be judged, should be ashamed to plead ignorance, if his life on earth has not been a useful and helpful one.

It pays to be informed. Knowledge is power: and success does not exceed power, and power does not exceed knowledge. They maintain an exact proportion. We do not farm for the sake of farming, nor are we in business for the sake of business. The ultimate end of all effort is character; the perfection of the man, mental, moral and physical. To wash, clothe, and feed the body, and for seventy-five years and daily repeat the process, while the mind is left to rags, filth and starvation, is to blunder fatally. The body for the brain, the brain for the mind, the mind for thought, and thought for action. The mind is the man.

In this age of newspapers and books a great mind can set the world moving in the direction of higher levels. As a consequence reforms are numerous, and the man who institutes and the man who completes may be the same, and no gray hairs or abated vigor will appear as witnesses against him to prove the long and wearisome length of the task he assumed; that which was once the work of centuries, is now accomplished throught he newspapers in a few years. But the majority of newspaper readers have come to know their value, and it is the intelligent use of them more than anything else, that needs to be emphasized. Murders, crimes of all kinds, removals, deaths, marriages, births, divorces, are news features of every successful daily or weekly; but to be satisfied if the paper contains nothing more, is to be harmed rather than helped by it.

From the newspapers we can learn how to take better care of the body, how to improve the mind, in what way the home may be made more attractive, the children more efficiently trained. It is this ability of the newspaper to answer and discuss every question that interests the thinking portion of the world that constitutes its chief value. Even a working man who will take the time and the pains to make a discriminate use of the newspaper, with the additional aid of a few books, may become fairly well educated. Carlysle says, the question is not, now-a-days, if a man has been through a university, but if a university has been through him: not where did you go to school, but what do you know.

The newspapers of Eaton County, eleven in number, have many intelligent readers. They should have more. It is difficult to understand how any citizen of the county can get along without taking at least one home paper. The adoption of proper methods in the conduct of the county's affairs, are due largely to the work of these papers in sending out information and presenting facts to their readers. When any question that concerns the future interests of the people is brought up, each voter's paper discusses it. The building of a new court house, the the construction of a bridge, the opening of a road, drainage, the care of the poor, the building of a school house, have all at different times been subjects for newspaper discussion and the information obtained by the people in this way has made a wise decision possible.

THE CHARLOTTE LEADER.

The Charlotte Leader was established in 1855 by C. C. Chatfield, at Eaton Rapids. It was then known as the Eaton County Argus. It was removed to Charlotte, with F. W. Highby as editor, in 1860. William Saunders became proprietor in 1861 and continued as such until his removal to California in 1865. From that date until 1868 it was published by D. F. Webber, who changed its name to the Charlotte Argus. J. V. Johnson then bought the office, but sold it again in 1875 to F. A. Ells, the present post master of Charlotte, by whom the name of the paper was changed to the Charlotte Leader. D. F. Webber is still a resident of Charlotte, a Justice of the Peace and has united to himself, by years of acquaintanceship and service, a host of esteemed friends.

Mr. Ells sold the paper in 1881 to W. G. Blymyer, who, on account of ill health, was compelled to retire, and was succeeded by U. C. Brandon in 1886. The Bryan Brothers have owned the paper since February 14th, 1888, and are well satisfied with the developments of the past and the evident prospects for the future. They were born on a farm near Fostoria, Hancock County, Ohio; Horton on January 9th, 1859, and Homer on August 17th, 1863. Soon after the war they were brought by their parents to Eaton County, where, with the exception of a few brief intervals, they have remained ever since. They learned the printer's trade in the office they now own, the elder entering it in 1875, and the younger in 1878. Horton did his first editorial work when the Prohibitionist was published in Charlotte, but later became city editor of the Ann Arbor Register. His single year's residence in the University City brought him valuable experience and aided in further preparing him for the management, in company with his brother, of the Charlotte Leader. May 12th, 1892, he was married to Miss Adele McClure, daughter of the late D. G. McClure. Homer K. Byran and Miss Edith, daughter of Mr. and Mrs. J. C. Haslett, of Charlotte, were united in marriage November 18th, 1890. Their home is brightened by the presence of two boys, Carl H. and Philip H. The Charlotte Leader, to which they are giving their best thoughts, is the only staunchly Democratic paper published in the county. Homer K. Bryan brought to the paper the knowlege gained by seven years' expepience on the metropolitan dailies of Chicago and is a practical job printer of much local fame. The Bryan Brothers are both Past Chancellor Commanders of Charlotte Lodge No 53 K. of P. and members of the Grand Lodge K. of P. of Michigan.

THE EATON COUNTY REPUBLICAN.

Has the following history: A paper that was started in Eaton Rapids in 1847, by L. W. McKinney, and afterward published by Dr. E. B. Burr, from whom it was purchased by Foote & Marsh, moved to Charlotte, and issued as the Eaton County Republican. E. A. Foote, the well known attorney, was editor, and Mark H. Marsh, a practical printer, afterward connected with the Evening News of Detroit, superintended the mechanical department. In 1859 Mr. Joseph Saunders became the proprietor and changed the name to the Charlotte Republican. Mr. Saunders was a man of large newspaper experience, keenly alive to new and improved methods, the first to use steam power in the printing business of the county, and the builder of a number of our substantial brick buildings. After a prosperous business of seventeen years he sold his paper to K. Kitteridge in 1866, who had been connected with several papers in the state, and is the present publisher of the Ann Arbor Register. In 1877 the Charlotte Republican became the property of Mr. D. B. Ainger. Mr. Ainger is now living in Lansing. From the first of May 1898, to April 1st of the same year, the Republican was edited and published by E. J. Tomlinson. Bissell & Jones are the present proprietors.

CHARLOTTE DAILY PRESS.

The newsboys' announcement of the Charlotte Daily Press is the latest evidence of Eaton County enterprise in the newspaper field. The Press is the first daily newspaper published in Eaton County. The initial number was issued from the Republican office, May 27th, 1895, and contained four pages of five columns each. L. P. Bissell and A. J. Munson were the promoters of the enterprise. L. P. Bissell was born in Medina County, Ohio, in 1854. His father was a Presbyterian clergyman, a graduate of several universities, and spent a number of years in educating his children in the languages and sciences. At an early age the subject of this sketch went into a printing office in an Illinois town and commenced his preparation for life as an adept in the art preservative. He made a study of the mechanical and professional branches of printing and publishing, and is well versed in almost every feature of the same, having worked in nearly all the large cities of the country in the various departments, printing, reporting, editing, etc. He is an earnest Republican, believing thoroughly in the principles of that party. He has only been in Eaton County for two years, but is rapidly coming to the front as an enterprising man. While engaged in the printing business in Ohio President Harrison appointed him post master of the little city where he resided. He is an alderman of the second ward of Charlotte, a Royal Arch and Council Mason, a member of the Republican Editorial Association of the state, and was unanimously chosen as secretary of the recent Congressional convention at Kalamazoo, which came so close to sending an Eaton County man to Congress. In 1887 he married Miss Frederika Salisbury, at Medina, Ohio. They have two children, Dorothy, and Paul Frederick, the latter born in Charlotte in December, 1898. Mr. Munson is a Michigan man, and has at different periods been engaged in newspaper publishing in this state. For the last few years he has been engaged in the different branches of newspaper work in Chicago, and brings to the Press a wide experience in metropolitan journalism.

THE CHARLOTTE TRIBUNE.

The Charlotte Tribune first appeared in August, 1887, with F. M. Potter as publisher. A half interest was purchased in December, 1887, by Geo. A. Perry, who was born July 8th, 1861, and is the eldest son of Mr. and Mrs. Horace B. Perry (see sketch) with whom he came to Brookfield, Feburary 8th, 1869, and has ever since resided in Eaton county. On September 19th, 1876, he married

Miss Belle McArthur, eldest daughter of George and Eliza McArthur, then of Brookfield. Their home is blessed with two daughters, Georgia Belle and Grace Avery, and another died in infancy. Mr. Perry's early life was devoted to farming summers and teaching winters, with an occasional term at Albion or Olivet colleges. At an early age he began to take an active part in politics, and he was elected supervisor of Brookfield for five consecutive terms. He resigned this office in 1882 to accept the county clerkship, which he held four years. While still county clerk he secured the release of two veterans from the poor house and through his instrumentality each was given a good pension. This was the beginning of a successful pension practice. On September 1st, 1889, Harry T. McGrath purchased a half interest and the firm name was changed from Perry & Potter to Perry & McGrath. In 1892 the the new firm moved into their fine new brick block from which the Tribune is now issued. Mr. Perry is identified with the best interests of the town and county and the Tribune, while Republican, is progressive, along all lines of reform. He is secretary and treasurer of the Eaton County Law and Order League to whose influence the recent increased majority for county prohibition is no doubt due. He is also serving his ninth consecutive term as secretary of the Eaton County Agricultural and Pioneer societies.

Mrs. Belle M. Perry, wife of Geo. A. Perry, conducts a valuable Woman's Department in the Tribune. She was for three years President of the Michigan Woman's Press Association and is now editor of the Interchange, the official organ of that association. The Interchange is printed at the Tribune office. Mrs. Perry is also President of the Charlotte High School Alumni Association, President of the Century Club, and was the first woman ever elected a trustee of the Charlotte schools. She has organized a fine club of Tribune writers, all of whom are members of her own sex. The Tribune is in good hands, taking their places. The Grange Visitor, the official organ of the State Grange, is published by Messrs. Perry & McGrath, who are its business managers. This paper is devoted to the interests of farmers, and has a large patronage throughout the state. Mr. Perry is a member of the Grange, the I. O. O. F. and of the Royal Arch Masonry.

THE EATON RAPIDS JOURNAL.

The Eaton Rapids Journal was founded by J. B. Teneyck in 1866, and sold to Frank C. Cully in 1869, who, in 1874, changed its name to the Saturday Journal. From 1876 to the 1st of January, 1879, Mr. E. O'Bien was the publisher. K. Kitteridge, his successor enlarged the paper and gave it the name under which it has since been issued. The present owner is C. T. Fairfield, who was born at Hillsdale, Mich., September 6th, 1866. He is the son of Hon. E. B. Fairfield, who was president of Hillsdale college for twenty-one years, Consul at Lyons, France, and for six years Chancellor of the State University at Lincoln, Nebraska. Here the son, C. T. Fairfield, was fitted for college. He entered Oberlin, Ohio, in 1884, and was graduated from there in 1887. He was financial manager of the Oberlin College paper for two years, and with this limited experience, and before attaining his majority he assumed control of the Eaton Rapids Journal. He is succeeding.

Chauncy W. Stevens, retired editor of the Eaton Rapids Journal, was born in Buffalo, N. Y., February 9th, 1825. He started in life for himself at the early age of ten years, and with a school career of but one year. His first venture was as a newsboy in the streets of New York City, hustling for "The Sun." He advanced with this paper from newsboy to roller boy, and finally compositor. In 1839 he came West with his parents and settled in Indiana. He entered the office of the Democrat at Goshen, and finished his trade under the instruction of Dr. W. H. Ellis. In 1856 he became proprietor of the Goshen Times, which he conducted for many years. During this period he was United States Marshal and United States District Enumerator of Elkhart, Indiana. In 1868 he sold the Times and purchased the Hudson (Michigan) Post. Three years later he located at Eaton Rapids and engaged in the manufacture of staves and heading. His business at that time was the largest manufacturing establishment in Eaton County. He purchased the Eaton Rapids Journal and successfully managed it for a time, when age and a competency bade him go out of business. Mr. Stevens is now well along in years and is retired from journalistic work, but his career has been a noble one. For forty years he has influenced public opinion and wrought out reforms.

THE EATON RAPIDS HERALD.

The Eaton Rapids Herald is edited and published by J. Dow Trimmer, a native of Ainger, this county, where he was born April, 21st, 1870. His parents moved to Charlotte while he was a babe. At the age of twelve Mr. Trimmer entered the office of the Luther Herald, at Luther, Lake County, this state, to learn the printer's trade. His winters, however, for several years were spent in school. A high school education and several years of practical experience have fitted him for the many positions he has held in connection with the printing business. He has worked on the Reed City Clarion, the State Democrat, of Cadillac, the Hudson Gazette and the Hudson Post, and was foreman of the printing department of the Central City Soap Company of Jackson for nearly two years prior to locating in Eaton Rapids, which was in March, 1894. Mr. Trimmer is a Democrat, and the Herald is independently Democratic. The Herald has had four or five different editors during its life of thirteen years. The present editor is a young man who is generally known as a hustler.

THE GRAND LEDGE INDEPENDENT.

Was established in January, 1869, by B. F. Saunders, son of Joseph Saunders, one of the earlier publishers of the Charlotte Republican. He induced W. C. Westland in February, 1894, to take a half interest in the business, and for a period of three years the paper was published by the firm of Saunders & Westland. The firm was dissolved the following May, the junior member becoming sole proprietor. For a time it was difficult to make the paper pay expenses. The business men and citizens, however, promised to give it their support, a promise which they have faithfully kept. Mr. Westland, who is still editor and publisher, enjoys the distinction of being the oldest in consecutive years of service in Eaton county, having edited and published the Independent for twenty-one years.

THE GRAND LEDGE REPUBLICAN.

We have been unable to obtain the details of this paper's history, but we have observed from its columns that it is a newsy sheet, with a good circulation. It's editor, M. H. Gunsenhouser is an able and experienced newspaper man. He was born in DeKalb County, Indiana, November 26th, 1854. He has worked in various cities of the country as job printer and editor. July 11th, 1889, Mr. Gunsenhouser, in company with M. J. Davis, purchased the Grand Ledge Graphic, a union labor paper. They changed its name to the Grand Ledge Republican, and likewise its politics were made Republican, for which party it is an effective worker. Fraternally Mr. Gunsenhouser is identified with the Knights of Pythias and Sons' of Veterans.

THE VERMONTVILLE ECHO.

Was started in 1871. It was called The Enterprise, but failed to become a financial success. Mr. Hawkins was the next owner, and F. M. Potter the next one. This purchaser gave the paper a new name The Hawk—but when it became the property of Holt & Knox it received its present name, to which it has clung ever since. After a few years Mr. Knox disposed of his interest in the business to John Sherman, who purchased Mr Holt's interest as half owner, and took possession of the office as proprietor and publisher. The firm is now J. C. Sherman & Son. John Sherman was born in Farfield, Franklin county, Vermont, October 6th, 1838. His parents, John and Persis Sherman, were natives Connecticut. From the district school in his native place, Mr. Sherman went to Bakersfield Academy, and the Academy at Rome, Vermont, where he received excellent training, to which his success as farmer and editor is due. He arrived in Michigan at the age of twenty and bought a farm which he still owns. In 1887 he began his career as publisher. Mr. Sherman was married to Miss Jane Boyce in March, 1855. She died in 1861. Mrs. Nellie M. Holt of Lansing, to whom Mr. Sherman was married in 1868, is the mother of W. E. Holt, editor of the Bellevue Gazette, and the partner to whom reference is made in this sketch.

THE SUNFIELD SUN

Was established by J. Q. Rounds, who continued its publication until November, 1891. His successor was I. N. Stevens, but his connection with the paper was severed January 23, 1895. Legge & Jenkins are now the proprietors, the purchase having been made February 27, 1895. Sunfield, the place of publication has a population of about five hundred. The paper is independent in politics devotes itself to the surrounding country. It has a circulation of about four hundred.

THE BELLEVUE GAZETTE.

Was established January 9th, 1871, by Mr. Alfred Linridge, who conducted it until May 29th, 1873, when it became the property of Edwin S. Hoskins. March 27, 1882, it was sold to G. W. Perry, in whose possession it remained for the next ten years. The present publisher of the Gazette is W. E. Holt, who was born in Canton, Wayne county, this state, June 4th, 1860. He was graduated from the Charlotte high schools in 1884, and on the 15th of October, 1895, he married Miss Lina V. Kennedy, a native of Vermontville.

Mr. Holt is a Republican and when election day comes is sure to be at the polls and equally sure to cast a straight ballot. When he edited the Vermontville Echo he served for three years in the capacity of village treasurer, recorder, one term and school inspector, two terms. He has been a member of the Ancient Order of United Workmen for a number of years, and during his residence at Vermontville, acted as recorder for that organization. He is also a prominent member of the Masonic Lodge at Bellevue.

THE OLIVET OPTIC

Was started in 1887 by Mrs. Stella Warner. During the first year of its issue it was sold to Fred Williams, whose proprietorship lasted but a brief time, and the present owner, Frank N. Green, purchased it. The Optic is a live paper, filled each week with news and sound editorials. It has a good circulation, is independent in politics, but is fearless in defending the moral side of all local questions of interest. The Optic has one of the best equipped offices in the county, from which are printed all the fine work, such as catalogues, programs, invitations, etc., of Olivet College; also the Echo, an illustrated magazine edited by the students of the College.

Mr. Green was born in Olivet in 1859. He attended the public schools and Olivet College until 1875 when he removed to a farm a few miles east of Olivet. He was on this farm for about eight years when he, returned to the village and purchased the Optic. This paper he continued to manage until November, 1894, when he was elected sheriff of Eaton county. He then placed his paper under the editorship of J. K. Swindt.

SCHOOLS OF EATON COUNTY.

By Rev. CHARLES McKENNEY, A. M., B. S.

ACKNOWLEDGMENTS: The author of this sketch wishes to publicly express his obligation to Hon. Frank A. Hooker, Hon. Daniel Strange, and Mr. George W. Sherwood for valuable information from which he has freely quoted.

On January 26, 1837, the last necessary step was taken by Congress, and Michigan became a state.

The population of the new state was 87,278, centered mostly in the southeastern portion. There were but few organized counties and Eaton was joined to Calhoun till 1837, and had in 1837 not to exceed 300 inhabitants.

It was the good fortune of Michigan that the foundation of her school system should be laid by such wise master-builders as Gen. Isaac E. Crary and Rev. John D. Pierce, both of Marshall. Mr. Pierce was the first to hold the office of Superintendent of Public Instruction, filling the position for five years, and to him more than any other, is due the credit for whatever worth and excellence our school system possesses. His general plan was to have a free public school in reach of every family, schools for academic training at favorable places in the state and a university for higher education. With some modifications the public school system of Michigan today is the same as planned by Father Pierce.

The early settlers of Eaton county were mostly from New York and states lying farther east, and brought into the western wilds the love of education which has ever characterized the people of New England and their descendants. They determined that the privations of pioneer life should not rob their children of at least a common school education, and soon after the scattered clearings, each with its rude fashioned log house, appeared the settlers joined in rolling up the log building which should be the home of the district school. In some instances, as in Oneida and Sunfield, a school was kept in a settler's home before the school house was built.

In 1835 the territorial council passed a bill organizing certain townships, one of which was Belleville (Bellevue) including all of Eaton county. The township now known as Bellevue contained about twenty-five families and here the first school in the county was taught by Hepzibeth Hutchinson in 1836. It is fitting that the name of this pioneer teacher should go on record and should be remembered by the never ending line of educators in Eaton county who, under more favorable conditions, shall carry on the work she so faithfully began.

The inspector's report for 1836 gives the valuation of school property at $70, the number of children, between five and sixteen, as fifty-six, and the total attendance seventy-two. For ten months school $64 were paid, the rate of wages being $1.50 per week. It is not stated that the teacher "boarded 'round" but we may safely conclude that the blessed experience was hers.

Gradually settlers began to locate in the other parts of the county, and in 1837 schools were taught in Eaton Rapids and Hamlin, though the townships were not then organized. A glance at table I will show how rapidly pioneers pushed into all parts of the county, and how the school teacher, the pioneer of education and culture, was abroad in the land to train the children who grew as plentiful and robust as the peach-blow potatoes, which the settlers planted between the roots of beech and oak stumps. Benton and Windsor were the last townships to support schools, but in 1843 schools were taught in every township in the county.

The limitations of pioneer life were well reflected in the school facilities of that early day. School house architecture had not reached the classical period. One common type prevailed. The school house was usually made of unhewn logs with the bark still adhering. The crevices were stopped by filling in split sticks and mud. Long shakes, split from ash or oak, covered the roof; the windows were few and small and the door turned on wooden hinges. A huge stick chimney often relieved the monotony of outline at one end. Knotholes in the logs often became the homes of yellow jackets and bees and not unfrequently snakes would be seen coiled on a projecting end enjoying a sun-bath till the recess or noon hour literally verified the prophecy that the seed of the woman shall bruise the serpent's head.

The exterior of the building was prophetic of what was to be found within. Rude logs formed the walls; planks hewn from logs with an ax made the floor. Soon, however, sawmills were started in the county and the ax-hewn planks gave way to the product of the mills. A huge fire place filled one end of the room and baked to a crisp those who sat near it while those farther away were shaking with the chill. On the sides, and at one end of the room, desks were made against the wall by boring holes into the logs, and driving in pegs on which boards were fastened. In front of these rude desks were benches made by splitting a log in halves and inserting legs under the convex side. The pupils while using these desks sat with their backs to the teacher and when classes were called they right-about-faced by lifting their feet over the benches on which they sat. In this position only did they have backs to their seats and they were formed by the sharp edges of the desk.

Books, whose names long since ceased to be heard in school circles, were studied by the barefoot lads and lassies of fifty years ago. Mitchell's geography, Adams' arithmetic, McGuffy's readers, Webster's and Townsend's spellers were the most common, and the attics of modern houses contain mutilated copies of these books that puzzled while they developed the brains of the children of the 40's. But while these books were common they by no means exhausted the list, for "uniformity of textbooks" was much farther away then than today. As a rule each child studied the books which chanced to be in use in the place whence he came, be the place New York, Massachusetts, Pennsylvania, Ohio, or Canada. Few branches were taught, and the three R's, "readin', writin' and 'rithmetic" with spelling, claimed almost the sole attention. Grammar was rarely studied and not till very recently did history, civil government, and physiology find a place in the district school curriculum. In very many instances the teachers were poorly qualified and methods were of the shake-roof and ash-bench order and yet, in the rude log school houses, poring over these ill-constructed books, guided, and often misguided, by the teachers, sat the boys and girls who, as men and women, have shaped the destiny, made the honorable present and made possible the more honorable future of Eaton county.

Educational enthusiasm which now finds vent in so many ways had its chief outlet in early days in the spelling school. Of the spelling school, Hon. Daniel Strange writes the following: "The spelling school was an important and frequent institution in Oneida for a quarter of a century. I have known a crew of young men to capture a pair of wild steers, never before yoked, and hitching them to a sled, push and pull them to the middle of the road, and with a whoop and hurrah, go merrily off to spelling school. If the sled tipped over or became entangled among the stumps there were enough on hand to right all again, and he was a luckless wight who failed to catch the sled when the steers again started on the run. Of course the more orthodox and satisfactory way was with a staid yoke of oxen and a long sleigh, with the box half filled with straw, but without other seats, the boys and girls sitting closely huddled, while practicing the spelling of such intricate words as 'necessary,' 'separate,' etc.

"You ask after the value of the spelling school. It was inestimable. It was the most important social institution of the day. It was there I found my wife. In an educational sense its price was above rubies. I knew a young man fitting for a spelling contest, who began spelling on Monday morning, and with different members of the family to pronounce words to him, he spelled till ten o'clock at night without missing a word. But he failed to win the contest, for though he had mastered four spelling books he had not learned all the words in the Geography. One who learned to spell at these contests has recently written a treatise of 200,000 words, and assures me that in writing the whole he consulted the dictionary for the spelling of but one word."

The following graphic account of the old-time spelling school is from the pen of Mr. Geo. W. Sherwood, who is well known throughout the county as a believer in the spelling school, and as a dangerous antagonist in the lists:

"What a multitude of reminiscences crowd thick and fast upon the memory as we contemplate the 'old-time spelling school.' Looming up in our mental vision stands the old log school house with its stick chimney, its long writing desk next to the wall, with oblong windows running parallel therewith, the big benches and the little benches, the long birchen whip on a couple of nails, just back of the teacher's desk in handy reach on short notice, and off in one corner that most dreaded and mortifying seat, 'the dunce block,' and last but not least the old-fashioned schoolmaster, with goose-quill pen over his left ear, rule in hand, pacing back and forth the room with all the assumed dignity of a country judge. All these old-time relics in panoramic view come up before me as I ruminate on this subject, relics of the long ago that this rising generation know but very little about. Being myself a student in those days, well do I remember with what joy and gladness every scholar hailed the announcement from the teacher that 'on Friday evening we will have a spelling school,' accompanied with the remark that the adjoining districts would probably 'catch on' and be out in force, and that we had better be prepared for them. With what renewed energy would the industrious scholar betake himself to the old 'elementary' in looking up all the hard words, if in so doing he might perchance be able to 'take down' the whole school.

"Night after night he may be seen by the dim light of his hickory torch poring over his spelling book until he becomes fully satisfied there is no word in that book he cannot spell. Again what a thrill of joy vibrates his every nerve as he contemplates the possibility of his being chosen by one of the captains in the coming contest, to sit beside the modest and fascinating blue-eyed Mary Jane, the prettiest girl in school, and thereby be afforded a most golden opportunity to engage her company home after school is out, much to the chagrin of that 'other fellow,' his most hated rival. But to my subject—the old-fashioned spelling school in Eaton County. It may not be amiss to mention right here, for the benefit of those who may never have attended one, the 'modus operandi' of an old-fashioned spelling school, which consists simply in extending a general invitation to meet at a certain time and place for a friendly contest in spelling. Having convened, the first thing in order is to elect a moderator, whose duty it is to keep tally of all the words lost and gained on either side, and, in conjunction with the teacher, to settle all difficulties that may arise in the contest. Next in order is the election of two captains, whose duty it is to choose sides; they generally draw cuts for first choice. Having chosen all that care to spell, the contest is now ready to commence. If any teachers are present from adjoining districts, out of courtesy they are generally invited to take part in pronouncing words. After spending about an hour in this exercise the teacher calls a halt and requests the moderator to report, which of course plainly shows which side has gained the victory. A recess of fifteen minutes is now taken, when the 'decks are cleared' for the final contest in 'spelling down.' Now is the time the average scholar feels a slight shock to his nervous system. Confidence, or lack of it, now plainly asserts itself. Having taken their places the teacher commences by pronouncing the first word of that old familiar lesson, 'all to be in trouble,'—and the trouble commences. One after another they go down like shocks of corn before the wind until there are but two left standing. The cross-firing now becomes fast and furious, and the feeling runs high as it is observed that one of these belongs to a neighboring district and is considered a champion of that school, while the other is of our own home school, and from the merry twinkle of his eye is plain to be seen that he feels himself master of the situation, and has 'come to stay.' A word finally strikes the former like a thunderbolt, when all eyes are centered on him as he hesitates, not knowing whether to say 'sion' or 'tion.' It is a critical moment, on which hangs possible victory or sure defeat. He finally guesses, and guesses correctly; his friends breathe a sigh of relief and the battle goes on. Back and forth the words fly in quick succession, when another one strikes the champion 'amidship' and down he goes all in a heap, and thus ends the contest in favor of the home school.

"It is with much regret that I learn on careful inquiry that for the last fifteen or twenty years the old-fashioned spelling school in the majority of districts in this county has been most sadly neglected. This is very lamentable, for nothing is of more importance in writing than good spelling. Poor penmanship may be excused, but poor spelling, in this age of free schools, never."

It is of interest in the light of Mr. Sherwood's closing lines to know that the Superintendent of Public Instruction is endeavoring to revive the interest in spelling by stimulating contests somewhat of the nature of the spelling school of early days.

Another educational institution, fruitful of unmeasured good, was the singing school. It gave play to social impulses, besides cultivating an art which added cheer to the home and interest and power to religious gatherings. During the winter months a singing teacher would organize schools in adjoining districts, to be held on different nights, thus occupying each night of the week. Merry loads of young people from one district would visit another and in not a few instances fires of love were kindled then that burn as warmly yet. The following sketch of the old time singing school is from the pen of Mr. Strange:

"If the spelling school was golden the singing school was solid silver, not yet demonetized. James Bailey, father of E. H. Bailey of Charlotte, taught our first singing schools. A great bulky lad of seven, I was sent along as escort to my elder sister. I remember well his quaint English accent as he counted 'wan, tu; wan, tu, sing.' I learned 'do' was on the added line, 'mi' on the first line, 'sol' upon the second line and so on for the eight notes. Imagine my surprise to learn later that 'do' meant the second line. I mastered this key with some courage, but when the following week 'do' took another skip, I refused to follow, and would sing only when the music was in my two keys.

"Do you ask the value of the pioneer singing school? 'By their fruits ye shall know them.' Previous to the singing school we had in our neighborhood two singers, my aunt Sally and a negro. When a hymn was read, they would consult together and he would whistle a tune half through, in a whisper, to see if it would fit. If it did not fit, he would try another until one in the right meter was found, when she would lead the tune and he would follow at a respectable distance. Young people were present but not one of them could sing. After the course of lessons by Mr. Bailey, a choir was organized that would do credit to any country church today. Our first chorister has since led the choirs in many villages and cities; our second has led a church choir in Lansing, and our third led the chappel singing at a Michigan college for three full years.

"One who had no instruction in music, beyond the Eaton county singing schools of thirty and forty years ago, has written sacred music which has met the approval of critics, and though the harmony is somewhat erratic, the wonder is, as Josh Billings said of woman's preaching and a dog's walking on two legs, that he could do it at all."

The writer, from actual acquaintance with the singing schools of twenty-five years ago, can endorse Mr. Strange's

Res. of James Baughman, Charlotte.

H—'s Woolen Mill, Eaton Rapids

S. T. Green, Implements, Charlotte.

Old Maid's Block, Charlotte.

Res. of D. D. Van Allen, Hamlin Twp. Portraits and Old Residence of Mr. and Mrs. Wm. B. Van Allen.

School Bldg., Grand Ledge, 1st Ward.

Sunfield Elevator, Sunfield.

Michigan State Bank, Eaton Rapids

Res. of Rev. W. R. Williams, Charlotte.

District No. 8, Walton Twp.

School Bldg., Grand Ledge, 2nd Ward.

Eaton Rapids City Schools.

Res. of Seth Ketcham, Charlotte.

OTTO MER.

PROPERTY OF W. B. OTTO, BENTON TOWNSHIP

OTTO MER is located in Benton Township, four miles north-east of Charlotte, and is the property of W. B. Otto. He began the breeding and raising of thoroughbred Perchron horses in 1880 and his first horse, the Noble Victor together with four brood mares, were secured from M. W. Dunham, of Wayne, Ill. From this time on he purchased quite frequently horses of this breed. The purchase price of Altino was $1,600.00. This horse was popularly known as Eaton County's favorite. Altino was sold in 1887 but his place was soon filled by Favoria, a $2000.00 stallion. The purchase at this time of the imported Constante and of the two famous black mares, Edith and Lauretta, was the laying of the foundation of his herd of beautiful black Perchrons. These three Perchrons are widely known as breeders of prize winners and among their productions may be mentioned the first pure bred black Perchron ever foaled in Eaton county—the greatest prize winner in the state of Michigan—the black beauty, Prince DeConde and the wonderful mare, Pride of Benton, noted as the first pure bred filly ever foaled in this county and she has also distinguished herself as a prize winner.

The accompanying illustration of this herd of black Perchrons shows Constante, a horse of 2000 pounds weight, who is known as a winner of first premiums at state and county fairs, Prince DeConde whose reputation as a magnificent animal has gone far beyond the limits of his county; Royal Star; a close follower in fame of his half brother, Prince DeConde; John L. Sullivan, a four year old who is justly classed with the rest of this noble herd; Edith, a three year old filly of excellent quality; and the black filly called Beauty, a three year old; and last but not least, the greatest prize winning mare of the state, the famous Edith with her filly colt Bessie at her side.

view of their importance. Modern methods in music select the few who may have special talent, and pass by the many of average musical ability, with the result that we have a few cultured singers, while the art that most of all adds to the home and to the social and religious gatherings is closed to the many. A revival of popular instruction in music would be a blessing.

A history of education in Eaton county should not fail to mention the Lyceum and Debating societies which, in many parts of the county, played no insignificant part in the training of young men in extemporaneous speaking. In some districts these clubs were maintained winter after winter with no loss of interest and zeal. The school house was the place of meeting and was often packed to its utmost, sleigh loads of visitors from adjacent districts often helping to swell the throng.

The questions debated did not vary much from year to year, and every winter the pros and cons of capital punishment, and the relative greatness of sword and pen were tested. The arguments of the veterans in debate came to be well known, yet seldom were tiresome, while each winter brought out some young knight ambitious to win his spurs in this verbal tournament. These societies served a good purpose. They developed social and intellectual life, and some of the most popular orators of Eaton county began their careers as speakers in the debating societies.

It has been a matter of congratulation and pride to the citizens of Eaton county that almost from the earliest settlement there have been within her borders institutions that afforded opportunities for at least academic education.

The colony which settled Vermontville, before leaving Vermont, had drawn up articles of agreement stating the purposes for which they were seeking homes in the West. The sixth article reads: "And we do also agree that for the benefit of our children and the rising generation, we will endeavor as far as possible, to carry with us and perpetuate among us the same literary privileges that we enjoy here." Amid all the hardships of pioneer life, they ceased not to remember the vow made in far off Vermont, and in 1842 began to carry it into effect by organizing the Vermontville Academical Association, which was incorporated in 1843 with nine trustees and a capital stock $10,000. A building was erected before the incorporation of the Association, and in the winter of '44 and '45 the academy was opened with Rev. W. U. Benedict, the pastor of the Congregational church, as teacher. The higher English branches and Latin were taught. The academy building still stands, serving as chapel to the Congregational church.

The academy was a beacon light of learning in the new land and attracted to itself students from not only the immediate vicinity, but from even remoter places. For twenty years or more the academy was sustained, but as colleges sprang up in southern Michigan, and union schools were organized in every considerable village, the demand for it ceased and in the later sixties it was merged into the Vermontville Union Schools.

But it had not lived in vain. It left an enduring impress upon Vermontville society and its influence was wide spread and gave vitality to the lives of many who are filling responsible positions with ability that only thorough training can give.

William Johnson established a high school in Charlotte in 1845 or 1846, and was himself the teacher. Mr. Foot writes of the institution and its associations as follows:

"From this germ of a school finally sprouted and took root the ambitious project of organizing a joint stock company and incorporating the Charlotte academy. Not only here, but in all of the adjoining townships, stock was liberally subscribed. Mr. McComb, a land owner, donated the academy ground and a bell. Mr. Nathan A. Johnson was the fortunate bidder who got the job of putting up this academy, and of collecting the subscription for his pay. The work was commenced about 1846, and for years was patiently prosecuted by Mr. Johnson alone. I remember of his starting to slide off the roof while shingling, and as he got to the eaves and was about shooting over, there happened to be a solitary staging pole about four feet away, against which he fortunately placed one foot and stopped his progress.

When Printer Johnson suspended his Eagle and went back to Ohio to reside, the academy enterprise languished, the subscription got cold, and Nathan found academy building up hill work, much more so than sliding down that roof. Just then people commenced finding fault because he was so slow in finishing the academy. I have a recollection of going down to that building one cold forenoon during the winter of 1848 and 1849, after it was roofed and sided up. I remember of climbing in, with no plank to walk upon, and of seeing windows boarded up with long boards to save spoiling the lumber. Joists, sleepers and studding were all bare. There were just boards enough for the work bench to stand upon in the middle of the lower room. All was silent save a long shaving hitched to a sliver by the side of a crevice, streaming and fluttering in the wind. There, all alone, sat the academy builder, Johnson, upon the work bench beside his empty nail box, his coat on and buttoned up to the chin, his purple hand clinging to the handle of his hammer. I couldn't see much sunshine in his countenance, and am sorry to say he found some fault because they were so slow in paying up their subscriptions. He had been dunning away at them for weeks without raising a dollar. 'Not a board,' said he, 'not a nail, not a sash nor a pane of glass, and those d—d fools all the while grumbling because I don't finish off this academy!' This was Wooden Johnson. Printer Johnson, the instigator of what led to this, had deserted us and gone back to Ohio.

Somewhere about 1850 the academy was finished; by what means I never learned. Several professors were at different times inveigled in there to teach, but found it up hill work in collecting their tuition. Professor Wallace stood it as long as he could and then went to railroading it out West and got killed. Professor Loring and his wife taught a few pupils, and boarded themselves frugally in a small room upstairs in the academy, until they starved out and went to farming over in Eaton, where the Professor soon died. Professor Ingham, who seemed expressly constituted for such usage, burrowed in this field of thorns and thistles until the organization of the Union school gave him a salary. He subsequently became a newspaper editor in Nebraska.

What dividends the stockholders received upon their subscriptions I have never learned. The stock was cheaply bought up by the few. The old academy has been sold and moved upon a front lot, and is now turned into the Peninsular Hotel. The entire ground has been cut up into city lots, and the academy is probably finished.

The Vermontville Academy was just opening when a company of thirty-eight persons, with goods in wagons drawn by ox teams, driving their flocks and herds before them, left Oberlin, Ohio, for Olivet, Michigan. The town existed only in name.

A few months previous, Rev. John Shipherd, the founder of Oberlin College, visited the Grand River valley, Michigan, to look after some land belonging to the Oberlin College. He had already determined to found a Christian college in Michigan, but had not decided upon a location. He chanced to visit the present site of Olivet, and becoming lost in the dense oak under brush which then covered the spot, stopped at a settler's home and was entertained over night. Receiving directions, he resumed his journey in the morning, but, to his surprise, soon found himself at the same spot where he became lost the night before. Again he started, and again after endeavoring to pick his way through the heavy underwood, he came upon the same eminence. He decided that this spot should be the site of his proposed college. The hill he called Olivet and the little brook which flowed at its base he named Kedron, but the new name has never supplanted the former one of Indian Creek. The Olivet Colony, after a two days' journey, arrived at their future home Saturday, February 24, 1844. Dwelling houses were built, and land cleared, and soon buildings for the college were completed. Father Shipherd fell a victim to deadly malaria at the beginning of the enterprise, but the colony did not despair. The first distinctively college building was burned before fully completed, but faith was not exhausted. In 1845 the legislature chartered Olivet Institute which continued until 1859, when a college charter was obtained. In 1894 Olivet College celebrated her fiftieth anniversary. During the half century, the value of her equipment had grown from nothing to $175,000. 10,000 young men and women have received instruction, and 350 had been graduated. Olivet College ranks among the best educational institutions of the west, and is a source of pride to the citizens of Eaton county.

For a quarter of a century the schools of Eaton County increased more in number than in efficiency. The wages paid offered no inducement for young men and women to educate themselves for the profession of teaching. School apparatus was limited to an ill-assorted lot of books in the hands of pupils, a square yard of blackboard, made of matched lumber, cubes of chalk an inch square, purchased in many instances by the children who used them, and erasers made by covering one side of a block of wood with sheepskin with the wool on. Occasionally charts illustrating penmanship adorned the walls, but were never used. The branches taught were, reading, writing, spelling, arithmetic and geography, and commonly grammar; occasionally a class in algebra would be found. History, physiology and civil government were practically unknown. Grammar was largely a girl's study, for they could attend school during the summer, while the boys, who, after the age of twelve, were kept at home to work, devoted the three or four months of the winter term to the three R's.

Until 1867 the licensing of teachers was done by a township board, consisting of the township board and two school inspectors. The board elected one of its members "visitor," whose duty it was to visit each school in the township at least once a term, to examine into the work of the teacher, and to test results by examining the pupils. If capable men should chance to be chosen, such a system would have yielded fair results, but too frequently such was not the case. Loose examinations were superintended, and unprogressive schools were the net product. Nor does this statement impeach the general intelligence and faithfulness of the officers of that day. It simply implies that a man who occasionally interests himself in educational affairs, and whose whole thought is given to other lines, cannot in the very nature of the case do efficient work in school supervision, which requires technical knowledge.

In 1867 the law creating the office of County Superintendent of Schools went into force and F. A. Hooker, a young lawyer, now a member of the Supreme bench of Michigan, was elected Superintendent. In reply to a question concerning the condition of the schools at the passage of the law and the work accomplished by him, Mr. Hooker writes:

"The law of 1867 providing for county superintendents of schools was a radical departure from existing conditions. Three school inspectors had previously granted certificates and established and altered the boundaries of districts. So far as I discovered they seldom did more, though occasionally a man would be found among them who visited schools. This, if I am not mistaken, was without compensation. The qualifications necessary to obtain a certificate differed in the various localities. In the villages and more advanced townships they were more than in others. In the townships, especially the newer ones, when log school houses abounded, the granting of certificates was largely a matter of expediency, and depended on the character of the school to be taught. They were usually secured after the school was secured and often, the wishes of the school board went farther than the attainments of the teacher.

"Methods of teaching were individual, of necessity, each teacher having his own. As a rule they were very primitive. One or two schools had maps, none had globes or other apparatus. My first attempt was to raise the standard of examinations. Manifestly a uniform standard was the result of a single examiner, but the result was consternation upon the part of the patrons and teachers. The first examination did not produce teachers enough to teach half of the schools, but by holding private examinations, and granting discretionary and short term certificates all were provided for the first summer, and the community settled down to peace and quiet. It had been badly disturbed and the office was very unpopular.

"By fall the examination showed the result of work on the part of the teachers, and each succeeding examination gave better results, though two years was too short a time to accomplish a very great improvement. I worked persistently to introduce maps, globes and a few other things but I was not able to accomplish much in this direction. During the two years I held it, I developed my energies to raising the standard of teachers, and felt that a marked improvement was discernable. That seemed to me the first step, and was a necessary foundation for other improvements in methods of instruction which would inevitably follow."

The work so well inaugurated by Mr. Hooker, was carried on by his successors, Superintendents Townsend, Evans and Sheap; but the office was unpopular throughout the state, and after eight years of trial it was abolished. In place of a County Superintendent, there was elected a superintendent for each township. Unity of plan at once disappeared and the work accomplished by county supervision was gradually undone. The utter failure of township supervision may be judged from the fact that in 1880 only fifteen schools in the county had prescribed courses of study, but nineteen were classified, and there were only ten that did not change teachers during the year.

A demand for better schools led, in 1881, to the creation of a County Board Board of School Examiners, to be composed of three members, whose duty it was to examine and license candidates; the secretary of the board should visit schools when occasion demanded. Prof. J. Estabrook, J. L. Wagner, and K. Kittridge were members of the first board. From that day to this the schools have gradually improved. By the law of 1889, the secretary of the board was to give his whole time to supervision with the title of County Secretary of Schools. It was the good fortune of Eaton County to have for secretary Mr. Orr Schurtz, whose efficiency and zeal made her schools second to none in the state. In 1894 Mr. Schurtz resigned to accept an important position in the schools of Grand Rapids, and was succeeded by J. L. Wagner, the present commissioner, who had been a member of the Examining Board since its creation. Under the supervision of Mr. Wagner the schools have been efficient and progressive. In 1891 the name of the office was changed from County Secretary to Commissioner of schools.

When Mr. Hooker became County Superintendent in 1867 he found the village and city schools in good condition, and steady improvement has marked their history to the present day. Gradually, the larger places leading, graded courses were adopted and to-day every village and city school is following a definite course of study. From an accompanying table it will be learned in what year the various schools began to graduate classes and the number graduated from each. For fifteen years the schools of Charlotte have ranked among the finest in the state, and the diplomas from the high school are accepted by the State University for all courses. The Eaton Rapids schools are also upon the university diploma list, though not for all courses. The youth of no county in Michigan have better educational opportunities than the youth of Eaton County. Her district schools are as good as any. From them it is but a step to excellent high schools, and when the high school is completed a college stands prepared to give the higher education.

It is safe to say that no county in Michigan of the population and age of Eaton has had such large influence in the educational affairs of the state. Her teachers have achieved distinction. Miss King, Superintendent of Charlotte schools from 1877 to 1881, is Professor of history in the Michigan Normal school; Miss Jones, Superintendent 1881 to 1885, is Professor of English Literature in the State Normal school of Kansas; Mr. Hammond, Superintendent from 1885 to 1893, is a member of the State Board of Education.

Eaton Rapids has been hardly less honored than Charlotte. Mr. Briggs, Mr. Schurtz, and Mr. Evans have all been called higher, and today hold responsible positions, as Superintendent of Coldwater schools, Principal of the West Side High school, Grand Rapids, and Superintendent of Jackson schools, respectively.

For twelve of the last thirty years Olivet College has furnished the state the Superintendent of Public Instruction, Prof. Hosford holding the position for eight years, from '65 to '73 and Prof. Estabrook four years, from '78 to '82. Mr. Schurtz, while Secretary of School in Eaton, was nominated by the republicans for the office of Superintendent of Public Instruction, but a political reverse lost the state to that party.

With honest pride the citizens of this county can look back over a third of a century and contemplate the material progress that has been made, and with even greater pride may they view the social, religious and educational advancement that have come with the years.

TABLE I. DATE OF THE FIRST TERM OF SCHOOL IN THE VARIOUS TOWNSHIPS.

Bellevue	1836
Hamlin	1837
Eaton Rapids	1837
Vermontville	1836
Eaton	1839
Walton	1839
Chester	1839
Oneida	1838
Kalamo	1840
Delta	1841
Charlotte	1841
Roxand	1841
Brookfield	1841
Carmel	1841
Sunfield	1842
Benton	1843
Windsor	1853

NOTE: The dates for Oneida, Windsor and Walton are not certain.

TABLE II. SCHOOL POPULATION BY DECADES.

1844	*150
1854	4,670
1864	6,314
1874	8,700
1884	9,351
1894	9,277

*School age in 1844 between five and seventeen. Other dates between five and twenty.

TABLE III. MISCELLANEOUS SCHOOL STATISTICS FOR 1895.

Children between 5 and 20	9,149
Value of School Property	$301,048
Male Teachers	74
Female Teachers	300
Total Number of Teachers	358
Wages Paid Male Teachers	$16,440.86
Wages Paid Female Teachers	$40,578.81
Total Wages Paid	$57,020.47
Primary School Fund	$14,978.33
Total Expenditures	$1,798.34
No. School Districts	148
No. Children Attending	7,453
Teachers required for Graded Schools	66
Ungraded	154
Average Price per Month (male)	$43.09
Female	$29.97
School House Material (brick)	43
Frame	118

NOTE: The last log school house ceased to be used in 1894.

Schools.	Grad's first class	No. Graduates Male	No. Graduates Female
Bellevue	[illegible]	[illegible]	[illegible]
Charlotte	[illegible]	[illegible]	[illegible]
Dimondale	[illegible]	[illegible]	[illegible]
Eaton Rapids	[illegible]	[illegible]	[illegible]
Grand Ledge No. 1	[illegible]	[illegible]	[illegible]
Grand Ledge No. 2	[illegible]	[illegible]	[illegible]
Olivet	[illegible]	[illegible]	[illegible]
Potterville	[illegible]	[illegible]	[illegible]
Vermontville	[illegible]	[illegible]	[illegible]

PHYSICIANS OF EATON COUNTY.

By WM. PARMENTER, A. M., M. D.

The practice of medicine in this state was regulated by no laws requiring study or preparation. The result has been that in the earlier history of this county the number of qualified physicians was few. In the early 40's it would have been possible to count all the doctors of this county on the fingers of two hands. Now there are more than fifty such physicians. Moreover, the requirements for graduation in these later days are three or four times as extensive as then. Medical schools were few, and the professional equipment of the young doctor was obtained in a course of reading in a doctor's office, attending at the same time to cleaning of spittoons and grooming of his preceptor's horse. In some cases he concluded his studies by a single or double course of lectures of four months' duration in a one-horse medical school; yet the majority went directly from their preceptor's office to assume the responsible duties of physician. The result was that a very few persons, endowed by nature with peculiar aptitude for the profession of medicine, rose superior to their surroundings and became noted physicians; noted more because of the contrast with the average physician of the time than because of actual attainments in knowledge and skill, compared with that of large numbers of our physicians of to-day; the great mass of doctors were contented, or compelled by their environments to be, and remained mediocres.

It must be said, however, that with the great advance of science in the last half century, medicine has not lagged behind. This is particularly true of preventive medicine and surgery. The causes of many diseases, especially contagious diseases, having been discovered, the means of their prevention was worked out and applied. The existence of disease producing germs in the atmosphere, and attached to all material objects, has called for and obtained the means for their destruction; and as a result, surgical operations are performed with impunity, which, before, were almost uniformly fatal. In the early 60's there were not a dozen surgeons in the United States who dared to open the abdominal cavity; now, in our own county, there are at least four who have successfully removed abdominal tumors, and did not think it "in form" to "sound a trumpet before them", on account of it. Thirty years ago the diseases in the autumn and early winter were almost exclusively malarial. Large quantities of quinine, at $2.50 to $4.50 an ounce, were prescribed, often amounting to an expense of the drug in one round of the physician. The diseases during the remainder of the year were largely the results of the malarial poisoning of the internal organs and nerve centers. This has changed. Ague is a rare disease and yet it lingers in the vicinity of sluggish streams and ponds. The practice of medicine then was attended with many inconveniences and hardships, among which were the sparsity and scattered condition of the population, requiring long rides on horseback over corduroy roads, and along bridle paths through the woods, to find in a log cabin whole families shivering with the ague, or pale and cadaverous, awaiting with dread the hour of the return of the swamp fiend; and he was sure to come. The scarcity of surgical supplies at that time is well illustrated in a very common incident occurring in the year 1858. It became necessary to amputate a torn or broken limb in the town of Vermontville, but no instruments could be found nearer than Eaton Rapids, 24 miles away. The physician in charge set the hour of operating at ten o'clock the next day, as it was then near evening; then mounted his horse and rode all night—forty-eight miles and was home in time to remove the limb without delay. It was his first amputation.

While the present physicians of this county compare favorably in ability and success with any others in the state, there are none who have attained special prominence, or who have been called to professional chairs in medical schools. One, who thirty years ago laid down the scalpel and medicine case, after ten years of service, has found in other fields a prominent place. Dr. R. C. Kedzie, a graduate of the first medical class in our State University, was called in 1863 to the chair of Chemistry in the State Agricultural College; and has since become prominent in sanitary and scientific circles. A very brief history of medical practice in the various localities in this county will be given in the alphabetical order of the localities. Under each local heading will be found the names and qualifications of each physician now in practice, as reported by himself. It will be noted that four ladies are now in the field. Twenty-two years ago there were none.

It has been attempted to give in these pages the name and address, at least, of every physician in the county. If anyone has been overlooked it has not been for want of careful and persistent inquiry.

BELLEVUE.

This is the earliest settled town in the county. A Dr. Carpenter was probably the first doctor who settled here, but of his history nothing can be obtained. Later came Drs. S. H. Gage, Fero, Taylor, Marshall and sixteen others, who remained but a short time. The present practitioners are A. W. Adams, Erastus Berry, A. S. Wilson and Horace D. Hall.

Dr. Albert W. Adams received his diploma from the Medical Department of Michigan University in 1875. He also took an *ad eundem* from the Bellevue Hospital Medical College of New York, the following year. He began his professional career in Kalamo, in 1878, but removed to Bellevue in 1882, and is in active practice there at the present time. He has never had time for office. He acquired the title of Ph. D. from the Michigan University and has since been satisfied with his titular honors. He is now in the prime of manhood.

Dr. Erastus Berry, now seventy-two years of age, commenced business as a physician at about forty-two. He graduated at the Detroit Medical College in the year 1871, six years after commencing his practice. Of late years, and at present, he has combined the business of druggist with the practice of medicine. The first year as druggist he took out a United States revenue permit to sell liquors, paying the usual fee of $25. Determined to sell according to law, and strictly for useful purposes only, he found at the end of the year his sales, all told, amounted to but $10, and said good bye to the liquor business. He is married, but without children.

Dr. H. D. Hall graduated at Hahnemann Medical College of Chicago, in 1870, at the age of forty-one. He is married, has a fine farm just outside of the corporation and has not aspired to political honors. He has been continuously the health officer for fifteen years, the entire time of his practice in Bellevue.

Dr. A. S. Wilson is a young physician of twenty-seven, a graduate of the State University in 1891, and has practiced only since graduation. Some future historian may be able to write a brilliant history of his future career. He is married and glad of it.

BROOKFIELD.

This town has been supplied with a physician most of the time since 1860. The order of their practice is: Dr. Thomas, 1860-1863; Dr. D. T. Williams, 1870-1887; Dr. W. H. Van Aude, 1863-1886; Dr. W. E. Newark, 1889-1891, and Dr. C. H. Sackett, the present incumbent, who came in 1894, a young married man from the Eclectic Medical Institute of Cincinnati, which gave him a diploma in June, 1894. He has a promising future.

CHARLOTTE.

The first doctor to attempt the task of visiting the sick in this city, then a little hamlet, and surrounding woods and marshes, came in 1842 but soon removed to Eaton Rapids, where he died. His name was Ralph. He was followed by Dr. J. P. Hall '49, who also died twenty years later. Contemporary with him was Dr. H. M. Manson who finished his work on earth ten years before the death of Hall. Dr. A. H. Sampson came in '53 and died in '65. Dr. Chas. A. Merritt came in 1854 and in 1867 Dr. G. T. Rand, a homoeopathist, located here but died in 1890. The present physicians in active practice are G. B. Allen, P. D. Patterson, A. B. Stealy, E. C. Palmer, F. A. Weaver, C. A. Merritt, W. E. Newark, Warren Rand, Sara J. Allen, H. J. Emery, and Mary K. Green.

Philo D. Patterson claims the State University as his Alma Mater, in 1868, and has used his knowledge and skill as a physician ever since that date; one year at Marshal, two at Carlisle in Kalamo township, and thirteen years in Charlotte. He is now fifty-one years of age, married. He filled the office of County Clerk for six years, from '73 to '78, has been First Vice President of the State Medical society, and has also held several other honorable positions in that body. Besides his degree, M. D., he holds that of B. S. conferred by Hillsdale College.

Dr. G. B. Allen has spent 35 years in Charlotte in the healing art, and now (1894) finds himself a member of the lower house of the State Legislature, his "first political offense." He received his degree of M. D. from the State University in 1867, and has practiced medicine twenty-eight years. He is genial, able, reliable.

Dr. Mary K. Green graduated in Philadelphia in 1868, and began practice in New York City. She became a member of the New York Medical Society, and, later, of the Medico Legal Society; the medical journals, both British and American, referred to her as being the first woman ever admitted to any medical society. She was elected Judge of Food Products at the Columbian Exposition. Makes a specialty of the study of foods, as to their scientific and nutritive value. Has practiced in Charlotte 21 years.

Dr. Horatio J. Emery took his medical degree from Queen's University, Ontario, in 1881, at the age of 35, and has spent ten of the eleven subsequent years in attempts to heal the sick. He combines the cognate of druggist with that of physician.

Dr. Allison B. Stealy found himself dubbed doctor of medicine by the Rush Medical College of Chicago, February 18th, 1886, at the age of 28. "Sensible to the last," he married and settled at Carlisle, in this county, remaining but one year, when he took up his abode in Charlotte. His early life was a struggle for existence; but by dint of farming, teaching school and studying at intervals he finally succeeded in obtaining his title of M. D. at Chicago, in 1886. In 1894 he took a post graduate course in Chicago.

Dr. Charles A. Merritt, at the ripe age of 70, remembers that his diploma, signed by Henry P. Tappan, D. D., L. L. D., University of Michigan, bears date of 1855. He combines agriculture with medicine, the former for gain, the latter *pro bono publico*. He has been Mayor of Charlotte one term, and for 20 years Superintendent of the Poor. He is now the oldest physician in Charlotte, having lived and practiced there since 1855—forty years.

Dr. Emery C. Palmer has practiced medicine in this county for twenty-five years and two years elsewhere. He first located in Pottersville, where he remained twenty-two years and more since that time he has resided in Charlotte. He claims for his *Alma Mater* the State University in 1870, and took a post graduate course in 1885 at Rush Medical College, Chicago, with private course in gynecology. He has also combined, a part of the time, and at present, the drug business with the duties of physician.

Dr. Sarah J. Allen took her medical degree from the Hahnemann Medical College, Chicago, in 1881. She has practiced in Chicago, Burlington, Ia., Battle Creek, Mich., and Charlotte. In 1885 she gave three months' special study to medical electricity, under Prof. Mills of Binghamton, N. Y.

Dr. Lewis A. Snell has given twenty years to the practice of homoeopathy in this state; at Lyons, Mason and Charlotte, where he is now located. He claims graduation from Detroit Homoeopathic College, session of 1871 and 1875; from the Homoeopathic Medical College of St. Louis, session of 1875 and 1876, and Chicago Homoeopathic Medical College, session of 1884 and and 1885.

Dr. Frank A. Weaver, aged 57, widower, commenced business fifteen years years ago at Chester, where he remained eleven years, going thence to Charlotte. He acquired his medical equipment at the Detroit College of Medicine, in 1881, after studying in the office of Dr. George E. Ranney, Lansing, 1878. He is at present the Secretary of the Eaton County Board of United States Examining Surgeons.

Dr. Wallace E. Newark possesses his credentials from the Toledo Medical College, Ohio, dated 1888. Seven years he has followed his chosen profession, and now, at the age of 36, finds himself installed in practice in the beautiful City of Charlotte. His first year was given to Nashville, Barry County; then five years to Brookfield, in this county.

CHESTER

Has had but three physicians since its foundations were laid. Dr. James L. Johnson first served that town from 1870 to 1882, then came Dr. F. A. Weaver, who was finally succeeded in 1893 by Dr. L. Frank Kies, a gentleman now twenty-nine years old, married, a graduate of Michigan University in 1893. He claims five years active practice, four of them in this county: first at Sunfield one year, and now three years in Chester.

DIMONDALE.

As early as the year 1845, Sylvester Derby located in the town of Windsor near the present site of Dimondale and treated the sick, made chairs and hunted wild game for a living. His knowledge of medicine was "picked up," yet he was considered a person well informed for the times. He was the first physician of the town, if tradition is not at fault. Then came Martin E. Manger, a graduate of the State University in 1865. The present physicians are Drs. Tyler Hall and E. S. Walford, both of Dimondale.

Dr. Tyler Hall has tried for twenty-eight years to cure the people in and about Dimondale. He began in the in the year '67 after having attended one full course of lectures at the State University. His graduation from the Detroit Medical College dates from 1871. He took a practitioner's course at Rush Medical College, Chicago, in '85. He studied law and was admitted to the bar nearly two years since. In 1887 he was elected to the lower house of the State Legislature.

Dr. Edgar S. Walford is thirty years of age, married, and has seen six years of practice. He located in Dimondale in 1889, one year after graduating from the College of Physicians and Surgeons, Chicago. He tried and abandoned the drug business, in connection with his practice.

EATON RAPIDS.

In the year 1838, so saith tradition, there came to this locality a Dr. Hart, but whenever he came or where and whither he went, this scribe has not been able to learn. Among those of later dates are found the names of Drs. C. N. Hayden, C. Metcalf, W. B. Hunt and A. C. Dutton, the last still resident but retired.

The present Medical faculty consists of Drs. Wilkins, Walker, Knight, Gallery, Bradley, Stimpson, Hyde, D. A. Long, Fred Long, Mary A. W. Williams and Henrietta A. Carr. Of all these, Drs. Wilkins and Knight are the earliest. A brief notice of each and all, whose history could be obtained, is given below.

Dr. Amos Knight, aged fifty-five, is contemporary with Dr. Wilkins; both locating here in the year 1866. He is a graduate of Cleveland Medical College, Cleveland, O., 1859; took an ad eundem degree in '71 at Rush Medical College, Chicago, and again at the College of Physicians and Surgeons, New York, in '75. He is married and has one child, a daughter.

Dr. Samuel M. Wilkins came to Eaton Rapids in 1866 and, with the exception of two years spent in Dakota, has continued in the healing art in that city now twenty-nine years. His degree was conferred by the Charity Hospital Medical College, Cleveland, O., in '66, and an ad eundem degree, in 1871, was bestowed by the Wooster University of the same city. He is nearing the sixties and has suffered for several years from poor health. He has been twice elected to the State Legislature, one term in each house; has served as

MEDICAL GROUP.

1 Sara J. Allen, M. D., Charlotte
2 G. B. Allen, M. D., Representative, Charlotte
3 Mary R. Green, M. D., Charlotte
4 L. E. Higbee, M. D., Pottervile
5 B. F. Willey, M. D., Sunfield
6 E. M. Snyder, M. D., Sunfield
7 O. S. Bailey, M. D., Hoytville
8 Hiram Walter, M. D., Eaton Rapids.
9 D. T. Williams, M. D., Brookfield
10 F. A. Weaver, M. D., Charlotte
11 W. R. Vanande, M. D., Sunfield
12 A. R. Stealy, M. D., Charlotte
13 M. S. Phillips, D. D. S., Charlotte
14 F. J. Stocking, D. D. S., Charlotte
15 Frank H. Honey, D. D. S., Charlotte
16 A. S. Wilson, M. D., Bellevue
17 P. D. Patterson, M. D., Charlotte
18 L. C. Jones, M. D., Kalamo
19 A. K. Warren, M. D., Olivet
20 C. Hooker Mead, M. D., Olivet
21 J. S. Newland, M. D., Olivet
22 Abram N. Hixson, M. D., Grand Ledge
23 A. C. Dutton, M. D., Eaton Rapids
24 Sam'l M. Wilkins, M. D., Eaton Rapids
25 Frank Merritt, M. D., Charlotte
26 R. S. Walford, M. D., Dimondale
27 Wm. Parmenter, A. M., M. D., Vermontville
28 Tyler Hull, M. D., Dimondale
29 W. A. Davis, M. D., Grand Ledge

Mayor of his city; and has held the position of local surgeon of the M. C. and L. S. & M. S. railroads. He served as a Union soldier in the late war till its close, in Reg. 102, Ohio Volunteer Infantry. He is married and has one child, a daughter.

Dr. Allen C. Dutton, a retired physician now seventy-one years of age, came to Eaton Rapids, June 10th, 1868, after eight years of medical practice elsewhere. He took his degree from the Western Reserve Medical College, Cleveland, O., in 1858. He withdrew from medical practice in 1881, since which time he has been druggist, merchant and banker. He has been called to, and filled, many local positions of usefulness and responsibility; among others, he has been member of the city school board for twenty years. He is a genial, high minded gentleman, with good prospect of many years of useful life.

Dr. David H. Long is an alumnus of Hahnemann Medical College of Chicago, 1870, and has devoted his attention to medicine at Eaton Rapids for fifteen years, giving to Ellendale, North Dakota the other two years of his professional life. He is, at the age of fifty-six, married and in active practice. He has had no hankering for office.

Charles Augustus Stimson M. D., a young and promising physician, has a diplome from Ann Arbor '91, and a marriage certificate, date not given. He made a special study of eye and ear diseases while at the University, yet he also engages in general practice.

Dr. Henrietta A. Carr, located in Eaton Rapids in 1893 coming direct from the University of Michigan, where she had just completed her medical studies. At twenty-five she is unmarried, and has been elected health officer of the city the past two years.

Dr. Mary A. Williams, aged forty-one, has done duty as a physician for four years in this county. Her present address is Eaton Rapids. Her diploma, from the State University, bears date 1891. She is the wife of Hon. W. W. Williams.

Dr. John M. Gallery, in the prime of life, has had eleven years' experience in medical practice, all in Eaton Rapids. His commission comes from Ann Arbor, '83. He has served on the Board of United States Pension Examiners, but has not attained political preferment, probably has not sought it.

Dr. Hiram Walter has attended sick calls for nearly fifteen years. In the year 1880 the Medical Faculty of the State University thought him qualified to prescribe for the sick, and so certified on a parchment of that date. He has held the office of Mayor of the city, and, for four years '81 to '85, was one of the United States Pension Examiners. He is in the prime of manhood.

GRAND LEDGE.

This city supports eight physicians, and is not considered very unhealthy either. No traditions of its early trials with its medical men can be obtained. Dr. Ambrose Brown and Dr. C. J. Corey are the earliest of those now resident in the city. Dr. Lamb was an old physician here in the fifties. Dr. S. S. Messenger came in 1859 and died in '89. He was an alumnus of the Michigan University Medical Department. Besides Dr. A. Brown, Dr. Davis and Dr. Dixson, who are sketched below, there are Drs. C. J. Corey, Geo. Green, W. R. Wilson, R. D. F. Brown, and Solon Whelpley, whose biographies this scribe has not been able to obtain.

Dr. William A. Davis came to Grand Ledge in 1870, after twelve years of hard service in Ingham county, Mich. At the age of twenty he commenced his studies at Ypsilanti preparatory to teaching, but abandoned that idea four years later; and, taking a three years' course of lectures, he graduated at the State University in 1858 with the title of M. D. He is now sixty-four years of age, married, and is "on call" night or day, especially day.

Dr. Ambrose Brown took his degree from Charity Hospital College, Cleveland, O. After thirty years of service at Grand Ledge, he is reminded that he is sixty-four years old. He served as Assistant Surgeon of the 193rd Ohio Volunteer Infantry in the late war. He has attended strictly to his profession since locating here.

Dr. Abram N. Dixson in mature life, has attended sick calls for seventeen years; twelve at Hoytville, and five at Grand Ledge, his present home. The State University gave him the title M. D., 1868. He is one of nine sons, but not the "seventh." He owns a farm of 160 acres in Roxand; and is, in consequence, well rated. His fine residence in the city is located in a garden where he worked for his board, while pursuing his medical studies.

HOYTVILLE.

This place, so named from Dr. Henry A. Hoyt, who first settled there and gathered around him a little cluster of houses and people, to whom he ministered as physician and merchant, has had but two doctors in its history—Hoyt and Bailey.

Dr. D. S. Bailey, in 1865, came here direct from the State University, and has remained ever since. He never had the political fever. His parentage was Scotch and English; born in Ontario, educated at Lansing and Ann Arbor, where he took special literary courses in connection with his medical studies. He has failed to marry.

KALAMO.

This town has at present two physicians—Dr. L. C. Jones, of the regular school, and Dr. Frank L. Snell, homeopathist.

"From the most ancient times" to the present this town can boast of twenty-two different physicians, most of whom have tried "their 'prentice han'" and then removed to fairer fields. Of these Dr. J. P. Comna was the pioneer. He is still living, and continuing his practice in Ohio. Dr. John Hall was a contemporary of Comna during the 40's, but has "passed over". He was located at Carlisle. Of the later practitioners, mention may be made of Samuel Perkey, now of Chicago, Joseph H. Griswold, a prominent physician of Grand Rapids, and J. H. Johnson, of Lisbon, North Dakota.

Dr. Lejeune C. Jones graduated from the Detroit Medical College, in 1885, from which college he later received ad eundem degree. He is thirty-two years old, has practiced ten years, the last seven in this county, and lives in single blessedness, after a few years of uncongenial married life.

Dr. F. L. Snell, a homoeopathic physician, has practiced the healing art for the last quarter of a century. He is a graduate of the Homoeopathic Medical College of Chicago, in 1886; is about fifty years of age and married. He has resided in Kalamo about sixteen years, but not continuously. Several of his brothers are also homoeopathic physicians in this county.

MULLIKEN.

Orlando A. Tucker, a young man and single, has tried the healing art for one year in this ambitious village. He comes equipped with papers from the from the Detroit Medical College, dated 1894. He was a student and practitioner in the office of Dr. Tyler Hall, Dimondale, previous to his graduation. His only competitor is Dr. G. V. Randall, who was the first doctor to locate here—about 1866.

Doctor Granger V. Randall has prescribed medicine according to the Hahnemann system for twelve years. The Chicago Homoeopathic College gave him his papers or sheepskin in 1883. He is the first who located at Mulliken.

OLIVET.

The first physician was Dr. Chas. Jennison, who came in 1845 and died in 1846. Then followed Dr. Chase, who also died a year later, and was succeeded by Dr. S. Kendall Orr, who continued to practice until 1855, and died, leaving a widow, who, in 1862, married Dr. Asa K. Warren, one of the three physicians now doing business in Olivet. Dr. Gordon, who succeeded Dr. Orr, died of small-pox in 1858. Drs. A. A. Thompson, Martin K. Munger, Asa K. Warren, Wm. Parmenter, Albert Thompson, Philip L. Green, Crane, Rusk, Hanes, Stockwell, C. H. Mead, M. L. Meads, George Weaver, Holmes and J. S. Newland have located here since 1856, in the order named, and have all removed or died except Warren, Mead and Newland.

Asa K. Warren is one of the oldest present practitioners of medicine in Eaton County. He graduated from Oberlin College in the year 1855, and in 1858 received his degree of M. D. from the University of Michigan. He settled in Olivet in the year 1859 and has been in almost continuous practice in that place ever since. His energy and force of character secured for him the respect and confidence of his fellow citizens, who elected him twice to the State Legislature (one term in each house) and continued him for fourteen years as their township supervisor, and, later, as county treasurer. He has owned and operated a farm in connection with his practice since 1863. Drs. Mead and Newland are at present his only competitors.

Dr. Chas. H. Mead is forty-seven years of age, is married and has one daughter. He studied medicine with Dr. Perkey, of Charlotte, dissected a subject in Carmel Hall; spent one summer in preparation with a Dr. Hooker, of Cayuga County, New York; then followed a year's course of lectures in the University of Michigan, after which he practiced in Olivet for eighteen months; then completed his studies in Chicago in 1878, and then came back to Olivet for a permanent location. For the past fifteen years he has combined the business of druggist with physician. He is now President of the Board of Examining Surgeons for pensions of the county.

Dr. J. S. Newland located in Olivet in 1885, after a practice of twenty years elsewhere; is a graduate of the College of Physicians and Surgeons of San Francisco, Cal., in 1877; he also graduated from the Eclectic Medical Institute of Cincinnati, Ohio, in 1883. He is a gentleman of about fifty, married and has not aspired to political honors.

POTTERVILLE.

Dr. E. C. Palmer first located here in 1870 and was followed by Dr. E. B. Eagle, in 1877. The present physicians are Drs. Eagle and Higbee.

Dr. Lewis S. Higbee has followed the profession of medicine about twenty years, nearly all of that time in Potterville. He is a gentleman in middle age, a graduate of the State University, in 1875, a brunette; has served the Chicago & Grand Trunk Railway for the past ten years as local surgeon. He has one competitor, in the person of Dr. E. B. Eagle.

Dr. E. B. Eagle, born in Moscow, Michigan, in 1855; was brought up as assistant to his father in Masherville in general merchandise; graduated in 1879, at the State University; married, as he should do; has located in Potterville, and, in addition to medicine, is a partner in a boot and shoe store. He ministers to the bodies and soles of his patrons.

ROXANA.

This little village supports two doctors at present—Armstrong and Lockrow. One physician preceded them—Dr. Abram DeCon—and one took the place of Dr. Armstrong after seven years—Dr. John J. Wester, since removed.

Dr. Thomas McKee Armstrong is forty-four years old, and began his practice in Roxana twenty-one years ago, continuing seven years at that place. He then returned to pursue his medical studies in the Detroit Medical College, and graduated in 1882, locating at Grand Ledge for seven years, he then went West, returning about one year ago to the old "camp ground." He was for a time a student in the office of Prof. R. W. Jenks, the President of the Detroit Medical College.

Dr. Calvin Lockrow is a homoeopathic doctor, claiming his medical equipment from the Hahnemann Medical College, Philadelphia in 1877, and has been in practice now eighteen years. He is slightly past a half century; married.

SUNFIELD.

This town and village claims at present four physicians—Drs. C. N. Snyder, E. M. Snyder, R. F. Willey and W. E. Vasmude. Early in the 40's Dr. R. F. Willey was found located on the East line of Sunfield as the first physician of whom the town could boast. Many years later (1877) came Dr. C. N. Snyder, located at Shaytown, and finally, on the founding of the village of Sunfield, about 1885, we find, in addition to the present physicians, about a half dozen others who remained but a short time. A brief sketch of the four resident doctors is given below.

Dr. Chas. N. Snyder is an alumnus of the Eclectic Medical College of Cincinnati, Ohio, in 1876. He began business at Shaytown in 1877 and removed to Sunfield Village in 1883, where he added the drug business to his practice, but has now abandoned that business. He is married, energetic and at forty-four gives promise of many years of hard work.

Dr. Edwin M. Snyder has prescribed for the sick of Sunfield for the past six years, commencing at the age of twenty-two, after getting his permit from the Michigan College of Medicine and Surgery of Detroit, in 1888. He also runs a drug store. He is married.

Dr. W. E. Vasmude graduated thirteen years ago from the Homoeopathic Department of the State University, and has done business as physician in this county since 1882, first in Brookfield and now in Sunfield; he confines his present business to office practice and the sale of drugs and medicines. He has lived nearly half a century and is somewhat broken in health. He is a registered pharmacist under the laws of this state.

Dr. R. F. Willey was born in Ohio, in 1820, and in early life studied in Oberlin, and later in a physician's office in Ohio. His college medical education was pursued at Ann Arbor and at the Pennsylvania Medical University of Philadelphia, where he received his diploma. He was the first to locate in the town of Sunfield. He has also practiced in Portland, Grand Rapids and in Kansas, and is now returned to the field of his former labors. He has a wife and one son.

VERMONTVILLE.

This village has, during its entire history, been visited by fifteen doctors. The first who undertook the unequal task of battling with the malaria and wild beasts came in 1837, but gave up the struggle with his life in one year. His headstone bears the name of Dr. Oliver Stiles. Dr. Dewey H. Robinson came in 1838 and was succeeded in a few years by Dr. Palmer. Then came a time when no physician lived here, and Bellevue, seventeen miles away, was the habitate of the nearest physician. This "interregnum" was followed by the advent of Dr. R. C. Kedzie, from 1850 to 1863. Dr. Stevens and Dr. A. A. Thompson were contemporary a part of the time with Kedzie, who was followed by Dr. Wm. Parmenter. There are now three resident physicians—Parmenter, Green and Snell. Their medical histories are briefly summarized below.

Dr. William Parmenter is third in length of practice in this county, being antedated only by Chas. Merritt, of Charlotte, and Dr. Asa K. Warren, of Olivet. He is sixty-six years old, the "husband of one wife." He received his degree of M. D. from the State University, in 1852, which was preceded by the literary degree of A. B., in 1854, from Oberlin College, Ohio, and followed, in 1857, by the honorary degree of A. M., from the same college. He spent the first four years after graduating in Iowa, returning to Olivet, Michigan, in 1861 and to Vermontville in 1863.

Dr. Philip L. Green is past middle life, has treated the sick for twenty-six years. He has been located at Vermontville for a quarter of a century. He is a graduate of Adelbert College, Cleveland, Ohio, in 1868. He took a post graduate course in Chicago in 1871, and again in New York in 1883. He has declined all office and has kept an eye single to medical practice.

Dr. C. S. Snell, a homoeopathic physician, located here sixteen years ago; a graduate of St. Louis Homoeopathic College of Missouri, in 1870. He had nine years experience in medicine in Whitefield, New Hampshire before coming to this locality. He is now in middle life, married, with a family of three children. He has been village President several terms and a member of the Board of Education a number of years.

ADDENDUM—CHARLOTTE.

Dr. James L. Johnston, at the age of 58, has retired from practice, after active service for thirty years. His office studies commenced in 1858. In 1863 he enlisted in the army and was detailed as assistant in the medical department, Eighth Ohio regiment, in which he remained till the close of the war. He assisted in the only successful case of re-section of the hip joint during the war, in March, 1864. He was the first doctor located at Chester, 1870-1883, when he moved to Charlotte, and retired from business in 1893, living now upon his farm, in that vicinity.

Dr. Warren Rand, homoeopathist, gives no auto-biography, yet he is in business in the city.

CHURCHES OF EATON COUNTY.

The larger log houses of the early settlers, with extemporaneous seats of boards and chairs, were the first places of worship in the county. When log school houses were built services were held in them. When the farms were cleared and the settlers had secured homes for themselves model houses of worship were built that cost no small amount of labor and self-denial. The ministers, who, between the years of 1835 and 1850, threaded their way through the forests of this county on foot or on horseback, fording the streams in order to reach their appointments many miles apart, have nearly all gone to their reward.

In 1833, at the house of Reuben Fitzgerald, in Bellevue, Rev. John D. Pierce, a Congregational minister from Marshall, delivered the first sermon ever heard in Eaton County. Mr. Pierce was the first Superintendent of Public Instruction in the State, and we are indebted to him more than to anyone else for shaping our excellent system of public schools, and also for outlining the works of the university. A full length oil painting of him adorns the walls of the reading room of Olivet College. In 1833 the Methodists organized in Bellevue the first church of the county. The second was organized by the Congregationalists in Vermontville, February 28, 1838, and was followed October 7, in the same year, by the Griffith M. E. church on the South line of Hamlin Township. In 1839 the Methodists of Eaton Rapids organized the fourth church in the county. A brief review of the several denominations of Eaton County is given below, but owing to a lack of data of some of them we are unable to give an extended notice.

CONGREGATIONAL CHURCH.

In 1831 a man named Joseph Dunton made his appearance in the county lecturing upon temperance and preaching. Some religious interests was awakened in the West part of the town of Carmel which resulted in the organization of a church with seven members, and it was called "The Congregational Church of Christ in Carmel." It met for worship in the Ells school house, situated a few rods North of the present Congregational church in Carmel. It had occasional preaching by Rev. E. N. Bartlett of Olivet College, and Rev. W. U. Benedict of Vermontville. In 1851 it changed its place of worship to the Court House in Charlotte, and its name to that of "First Congregational Church of Charlotte. On New Year's Day, 1854, Rev. Wolcott B. Williams became the first resident pastor of the church and continued for thirteen years. In 1856 the society erected, at the South end of Bostwick avenue, a frame house of worship at an expense of $1,775, for house and lot. It had two hundred and fifty sittings and was innocent of vestibule, bell and steeple. By outsiders it was dubbed "The Dew Wood Church." It was, however, the first house of worship in Charlotte, and at the time there were no others within ten miles. In 1874 the society erected its present house of worship which was not fully completed until 1881.

The Eaton Rapids Congregational church was organized July 14, 1843, by Rev. Joseph W. Smith, who was pastor for only about two years, but lived in Eaton Rapids until his death in 1867. After the close of his pastorate, for about two years, the church had only occasional services. In 1851 Rev. John S. Kedzie became pastor, and the church has ever since steadily maintained its church life. It occupied a small house of worship of its own for the first time October 30, 1846. It stood about three blocks South of its present location. In 1860 the building was removed to the present site and greatly enlarged and improved, but on the night of January 6, 1877, it was burned down and the present neat brick edifice was erected on the ground where it stood.

The Congregational church in Olivet was organized with seventeen members by a council that convened at the house of George Andrus, Olivet, March 26, 1843. At first its services were held in private houses. When the first college building was erected the upper story was fitted up for a chapel and the church worshipped there. The first place of worship was in a chapel across the East end of Colonial hall, the building now used for a gymnasium. In 1852 it united with the college in building a house of worship, which, in 1865, was enlarged by adding several feet to the length and putting under the whole a basement story. In 1894 college and church united in the erection of a beautiful house of worship, of field stone, containing some 900 sittings.

The Congregational church of Vermontville was organized February 28, 1838, by Rev. Sylvester Cochrane, one of the original colonists from Vermont, who preached for a time in the large log house of S. S. Church. When a log school house was built the church met there for worship. In 1844 a two-story frame building was erected for an academy, and the upper story was fitted up for a chapel and is still used as such. The present house of worship was built in 1862, during the pastorate of Rev. O. H. Spoor. During the pastorate of Rev. David Braton, in 1886, a lot was purchased and a commodious parsonage erected upon it at a total cost of $2,000. It is the oldest Congregational church in the county, and was the second such organized in the county. It has in fifty-six years had thirteen different ministers.

The Congregational church of Grand Ledge was not organized until 1861. Rev. Wm. P. Esler was the organizer, and the church had four original members. Its house of worship was erected in 1865 with two hundred sittings and at a cost of $2,000. It now has fifty-one members and one hundred and sixteen in Sabbath school. Its current expenses are $850 per year. There are at present fourteen churches with 1,853 members; their current expenses are $16,333. They give annually for benevolences $1,857. The value of Congregational church buildings in Eaton county is $84,800 and their combined seating capacity is 5,425.

CHARLOTTE METHODIST EPISCOPAL.

In 1836, Rev. E. H. Pilcher was Presiding Elder, and with the exception of Bellevue the Methodist work in Eaton county was known as the Ingham Mission belonging to the Marshal District. This Mission embraced also the west half of Ingham county. Rev. Washington Jackson was preacher and conducted services at the house of Jonathan Searles, two miles southeast of Charlotte. In 1839 Rev. Isaac Bennett was preacher and held services in the new court room in the hewn log tavern known as the Eagle Hotel that stood where the Phenix now stands. It is supposed that he performed the first marriage ceremony in Charlotte, the parties being Rollo Cushing and Susan Searles. In 1841 Rev. Levi Warner was preacher and the meetings were transferred to a hewn log building 18x30 feet that stood just east of the Sherwood House, and which, according to the exigencies of the times, did service as a school house, printing office, chapel and dwelling. In that year a class was organized with David Darwin Hughes as leader, who later became a distinguished lawyer. The next place of meeting was in a block school house on the corner west of Munger & Jenning's store. In 1845 the court house was built and for some thirteen years the people met there for worship. In 1859 the society erected a brick house for worship 47x80 feet, which is the main body of the present edifice. During the pastorate of Rev. C. S. Fox in 1873, this building was enlarged by putting a transept across the south end and carrying out a vestibule and steeple in front.

EATON RAPIDS.

In 1842 the Ingham Mission of the Methodist Episcopal church embraced the west half of Ingham county and all of Eaton county except Bellevue. Rev. Isaac Bennett was pastor in charge and organized a class in Eaton Rapids with nine members. During the pastorate of Rev. W. E. Bigelow in 1843, the church erected a small frame building for a house of worship. In 1856 Eaton Rapids was made a station with Rev. Richard Peagelly as pastor. In 1888 the society erected its present beautiful and commodious brick edifice.

GRAND LEDGE.

The first Methodist meeting in Grand Ledge was held in 1851 by the Rev. John Clayton, an earnest local preacher who is still living. The meeting was at the house of one Adam Smith. In 1860 the present Grand Ledge charge belonged to the Ionia Circuit, and Revs. O. Whitmore and A. R. Bartlett were preachers. No record shows the number in class at that date. Its house of worship, built of brick, is furnished with a town clock and is the only church in the county that is thus equipped.

VERMONTVILLE.

What is now known as the Vermontville M. E. church was originally in the same circuit as Eaton Rapids, and later with that of Charlotte. In 1859 Vermontville first appears in the minutes as a separate charge and Rev. Josiah Fowler was preacher. At the close of the year he reported one hundred and fifty-nine members and fifteen probations, but the circuit then embraced classes in Kalamo, Vermontville, Bismark and the Dow neighborhood. In 1865 Rev. Noble was preacher and organized the Vermontville class, which held its meetings four miles north-east of Vermontville, where they erected a house of worship during the pastorate of Rev. Thomas Clark in 1863. In 1871 Rev. J. W. Smith was pastor and moved the house of worship to the village of Vermontville. Rev. J. H. Thomas was pastor in 1887 in which year the church was enlarged by the addition of a wing, and was refitted. These improvements were made at a cost of $1,200.

HAMLIN.

The Griffith M. E. church in the southern part of Hamlin, although a country church in a remote part of the county, and but little known, is really one of the oldest religious organizations in the county. The first religious meetings held in the neighborhood were conducted by local preachers at the House of Stephen Reynolds. Those were days of boundless hospitality and Mr. Reynolds seems to have kept a veritable ministers'' tavern. At one time he had three local preachers and their families living in his house, viz: Revs. Wm. Crane, Truman Harrand and Turner. It is said that thirty-two persons lodged at Mr. Reynolds' at one time. It is not strange that such a household should have felt the need of religious services at least once a week nor that, with so many preachers on hand, Mr. Reynolds should have been able to secure them. In 1835 the Michigan Conference was held in Mansfield, Ohio, and sent a missionary into these parts by the name of Kinnear. In 1836 Conference met in Detroit and sent as missionary, Rev. Washington Jackson, who moved into his field with his family and lived in the house with the Hodgetts where three families were living before their arrival. Mr. Jackson preached at different houses in the Griffith neighborhood once in two weeks alternately with Grand Rapids some eighty miles distant, going to his appointments on foot, and holding meetings at intermediate points on the way. His first station west of Griffith's seems to have been the house of Jonathan Searles two miles south-east of Charlotte. In 1838 Conference sent David Thomas as preacher, who, on the 7th of October in that year, organized a class of fourteen members, and about the same time another class of seven members was organized at Truman Hart's; these two classes were soon after merged into one. In 1840 meetings were held in a school house. In February, 1855, the society had the pleasure of worshiping in their own new sanctuary. This church has had the benefit of the labors of twenty-eight different ministers besides those of local preachers and Presiding Elders.

BAPTIST CHURCHES.

In 1841 Mrs. Esther Searles, Julia Pierce and Mary Rager were, so far as known, the only Baptists living in the vicinity of Charlotte, and in 1842 they secured a visit from Elder John Tompkins, who preached at the old log building in Carmel known as the Ells school house, and then and there these three women were recognized as the Baptist Conference of Carmel. In 1843 they changed their place of meeting to the Court House in Charlotte and seven new members were added to their number, and these ten were recognized as the First Baptist church in Charlotte and the services of Elder Tompkins were secured for one-half the time. For several years services were held in the Court House, the Congregational church and in Carmel Hall. In 1869 a model frame building was erected for a house of worship. In 1882 it was enlarged, veneered with brick and a spire added, and more recently it has been still further improved and decorated.

On the 23d of February, 1843 Calvin Rose, Ann Rose, Henry K. Jeffries, Mary L. Jeffries, Timothy Wheeler, Asenath Wheeler, Ann Arnold and Samuel Ferris met and organized themselves into the First Baptist Church of Eaton Rapids, by adopting articles of faith and entering into covenant with each other. On the 17th of the following April it was formally recognized by council. For three years the church was supplied by different ministers, among them Rev. J. C. Post and Elder Hill. The first building occupied by the church is the present upright part of the old Dr. S. M. Wilkins house next to the present edifice. In 1856, under the pastorate of Rev. H. G. Meaher, a frame meeting house was erected on the present site. This was extensively repaired in 1876 under the supervision of Rev. J. M. Titterington. This was again enlarged and remodeled in 1890 under the pastorate of Rev. J. P. Feroney.

The Free Will Baptist church of Grand Ledge was organized in the Johnson settlement four miles west of Grand Ledge by Rev. S. A. Causey in 1851 (one authority says 1855). There was also a small society of the same faith in Eagle, Ionia County, and in 1871 these united and changed their place of meeting to Grand Ledge.

In the year 1883 a number of representative men from the Free Will Baptist and also the regular Baptist churches met in Jackson to see if it was feasible to adopt some plan to secure a closer union between the two denominations. After a protracted and friendly discussion an agreement was reached on the following principles, which are known as the Jackson Platform, and which is gradually being accepted by the churches of each body.

THE JACKSON PLATFORM

I. No abandonment of doctrines now held is required, but the regular Baptists are asked to simplify their acceptance of the doctrines of freedom and responsibility of man, and the Free Will Baptists are asked to simplify their acceptance of the doctrines of grace—the exact interpretation of all doctrines being left to the individual, subject only to the mind of the Lord as revealed in the New Testament.

II. According to the teaching and practice of the apostles, baptism should be received immediately after regeneration, and therefore properly precedes the Lord's Supper and public Christian duties.

III. It is advisable that invitations to the Lord's Supper be omitted, the announcement of the observance of the supper being considered sufficient.

IV. All our churches are advised to omit the use of the term "Sacrament," as defining the nature and meaning of the ordinances, and that "watchcare membership" be discontinued where it has existed.

V. It is suggested that our churches discontinue the use of the distinctive terms "Regular," "Particular," "Free" and "Free Will," and use only the name "Baptist."

VI. It is suggested that the churches of both bodies mutually accept church letters of commendation and ministerial credentials.

Inasmuch as there is a coming together of the Regular and the Free Will Baptists on the basis of the Jackson Platform we have included the Grand Ledge church among other Baptist churches in the county.

UNITED BRETHREN CHURCHES.

The first United Brethren church of Eaton County was organized in Walton by Rev. Koss. It had five members. In 1877 they erected a new house of worship at a cost of $800. It has a seating capacity of two hundred persons. They also have a $600 parsonage. In 1866 they organized a United Brethren society in Charlotte. Their present house of worship is a neat brick structure, erected in 1874, at a cost of $3,000.

In 1878 the United Brethren attempted to build a small brick church in Potterville, and laid the foundation and gathered some materials, but the leader became discouraged and called on Rev. W. B. Williams, then Superintendent of Missions, and requested him to organize a Congregational church and take the foundation and materials on the ground and go forward and complete the edifice. Mr. Williams sent word to Potterville that he would preach there on the following Sabbath. At the close of the sermon he stated to the congregation the request he had received, and inquired if anyone present had any objections to the transfer. No objection was made, and soon he organized a church of seven members raised funds to finish the house.

In 1880 the United Brethren erected a brick house of worship in Mulliken, but before it was paid for dissensions arose and the building was advertised for sale to pay debts incurred in its erection, and Rev. Leroy Warren, then Superintendent of Missions, was invited to organize a Congregational Church in Mulliken, and take the meeting house on condition of settling all claims against the property. This he did, organizing a church with eleven members, and securing a liberal grant from the Congregational Church Building Society and becoming personally responsible for the balance of the claims.

There are at present eleven United Brethren churches in Eaton County, with a total membership of three hundred and thirty-four. The total value of church buildings is $18,870.

CATHOLIC.

In 1851 there were, so far as known, not more than a half dozen Catholic families in Eaton county, and only two or three in Charlotte, and for several years thereafter no services were held in Charlotte, but during the time Patrick McDonald and wife went once a year to attend worship in Marshal, and finally induced Rev. P. C. Koopmans to visit Charlotte occasionally and celebrate mass in their private residence. The building of the Chicago & Grand Trunk railroad brought many more Catholics to the town, and in 1868 Rev. C. M. Tonia, with the assistance of Mr. McDonald, raised money to erect upon a lot donated by the priest, and situated on the high ground in the northern outskirts of the city, a modest edifice costing $1000 with a seating capacity of two hundred. In this effort they were generously aided by the Protestants. In 1871 Rev. F. Brugger, of Hastings was in charge of the parish and raised money to purchase a site upon which the present house of worship was erected in 1892 through the agency of Rev. P. Laughoen. In this effort also they gratefully acknowledge the liberal aid of Protestants.

In the spring of 1891, the Catholics of Eaton Rapids, under the lead of Rev. Francis Brugger of Hastings, bought the property of the Episcopal society in Eaton Rapids, put it in repair and furnished it, and in September of that year it was consecrated by Bishop Foley of Detroit.

EPISCOPALIAN CHURCHES.

In 1848 the Rev. L. Foote began holding the services of the Protestant Episcopal Church in the old Court House in Charlotte, and continued them for six years. Then for twenty years there were only occasional services. In 1871 a mission was organized under the name of St. John's Mission of Charlotte, subject to the visitation of Bishop Samuel A. McCoskry, and occasional services were held by Rev. G. P. Schetky and Rev. G. E. Peters. In 1873 a wing of the old Court House was bought and fitted up for a chapel. The Rev. J. L. Taylor took charge of the mission as permanent Rector in the Spring of 1874 and continued services in Charlotte and Eaton Rapids four years. In 1877 Bishop Gillespie changed the name of the mission to that of "Grace Mission, Charlotte." On the retirement of Mr. Taylor, services were suspended for several years. In 1887 they were renewed by Rev. J. W. Bancroft and for two years continued by him and Mr. Sharpe, a lay reader. In 1889 the present house of worship was built and Rev. Van Antwerp served as Rector for one year. Rev. M. H. Martin served as Rector from 1891 to 1893.

THE GERMAN BRETHREN.

This church, which is located at Sunfield, sometimes known as the Dunkard church, is the only one of this denomination in the county. One of its peculiarities is, they think it wrong to pay their ministers a definite salary. During the past year the church has made several contributions for benevolent objects of which no record has been kept. The denomination supports missions in Canada, India, Denmark, Sweden and Germany.

UNIVERSALIST CHURCHES.

Ever since 1851 there has been occasional liberal preaching in Charlotte by Unitarian and Universalist clergymen. For some months Rev. J. Parker, a Unitarian minister, held services here regularly. In 1870 a Universalist Society was organized. Rev. James Gorton was the first resident Universalist minister. He began his ministry in Charlotte in 1881. In 1880 the society was reorganized with seventy members.

FIRST HOUSES OF WORSHIP ERECTED.

The Congregational society in Olivet in 1852, completed the first house of worship in the county. The second one was build in 1853 by the Methodists at Bellevue. In February, 1855, the Griffith M. E. Church was dedicated and in the same year the Congregationalists and also the Methodists in Eaton Rapids completed houses of worship that they had begun in 1848. The old church, built in 1856 by the Congregationalists, in Charlotte, was the sixth meeting house built in the county.

The Township of Sunfield is better supplied with churches than any other township in the county. It has twelve church organizations and ten houses of worship. The Methodists are better supplied with parsonages than any denomination in the county. They have one for every charge.

The following tables will show the several historical points of interest concerning the remainder of the churches of the county.

SEVENTH DAY ADVENTIST.

Churches.	By Whom Organized.	Date of Organization.	Original Members.	Present Membership.	Annual Current Expenses.	Benevolences.	Houses of Worship.	Sittings.	Value of Church Property.	Value of Parsonages.
Charlotte	Rev. J. N. Loughborough	[illegible]	[illegible]	[illegible]	[illegible]	[illegible]	[illegible]	[illegible]	[illegible]	[illegible]
Eaton Rapids	Rev. J. O. Corliss	[illegible]	[illegible]	[illegible]	[illegible]	[illegible]	[illegible]	[illegible]	[illegible]	[illegible]
Potterville	Rev. Joseph Frisby	[illegible]	[illegible]	[illegible]	[illegible]	[illegible]	[illegible]	[illegible]	[illegible]	[illegible]
[illegible]	Rev. John Sisley	[illegible]	[illegible]	[illegible]	[illegible]	[illegible]	[illegible]	[illegible]	[illegible]	[illegible]
Vermontville	Rev. J. D. Van Horn	[illegible]	[illegible]	[illegible]	[illegible]	[illegible]	[illegible]	[illegible]	[illegible]	[illegible]
[illegible]	Rev. Franklin [illegible]	[illegible]	[illegible]	[illegible]	[illegible]	[illegible]	[illegible]	[illegible]	[illegible]	[illegible]
Total			[illegible]	[illegible]	[illegible]	[illegible]	[illegible]	[illegible]	[illegible]	[illegible]

PRESBYTERIAN.

Churches.	By Whom Organized.	Date of Organization.	Original Members.	Present Membership.	Annual Current Expenses.	Benevolences.	Houses of Worship.	Sittings.	Value of Church Property.	Value of Parsonages.
Oneida	Rev. William T. Benedict	[illegible]	[illegible]	[illegible]	[illegible]	[illegible]	[illegible]	[illegible]	[illegible]	[illegible]
[illegible]	Rev. William C. Benedict	[illegible]	[illegible]	[illegible]	[illegible]	[illegible]	[illegible]	[illegible]	[illegible]	[illegible]
Windsor	Rev. Calvin Clark	[illegible]	[illegible]	[illegible]	[illegible]	[illegible]	[illegible]	[illegible]	[illegible]	[illegible]
Total			[illegible]	[illegible]	[illegible]	[illegible]	[illegible]	[illegible]	[illegible]	[illegible]

METHODIST PROTESTANT.

Churches.	By Whom Organized.	Date of Organization.	Original Members.	Present Membership.	Annual Current Expenses.	Benevolences.	Houses of Worship.	Sittings.	Value of Church Property.	Value of Parsonages.
Ainger	Revs. A. Byers and S. Philips	[illegible]	[illegible]	[illegible]	[illegible]	[illegible]	[illegible]	[illegible]	[illegible]	[illegible]
Carmel	Rev. George A. [illegible]	[illegible]	[illegible]	[illegible]	[illegible]	[illegible]	[illegible]	[illegible]	[illegible]	[illegible]
[illegible]	Rev. L. F. Hall	[illegible]	[illegible]	[illegible]	[illegible]	[illegible]	[illegible]	[illegible]	[illegible]	[illegible]
Eaton Rapids	Revs. L. Dobbs and F. A. Perry	[illegible]	[illegible]	[illegible]	[illegible]	[illegible]	[illegible]	[illegible]	[illegible]	[illegible]
Total			[illegible]	[illegible]	[illegible]	[illegible]	[illegible]	[illegible]	[illegible]	[illegible]

CHURCH SUMMARY.

Denominations.	No. Churches.	No. Church Members.	No. Sabbath Schools.	Annual Current Expenses.	Benevolences.	Houses of Worship.	Sittings.	Value of Church Property.	No. Parsonages.	Value of Parsonages.
Universalist	[illegible]	[illegible]	[illegible]	[illegible]	[illegible]	[illegible]	[illegible]	[illegible]	[illegible]	[illegible]
Episcopalian	[illegible]	[illegible]	[illegible]	[illegible]	[illegible]	[illegible]	[illegible]	[illegible]	[illegible]	[illegible]
German Baptist Brethren	[illegible]	[illegible]	[illegible]	[illegible]	[illegible]	[illegible]	[illegible]	[illegible]	[illegible]	[illegible]
Catholic	[illegible]	[illegible]	[illegible]	[illegible]	[illegible]	[illegible]	[illegible]	[illegible]	[illegible]	[illegible]
Free Methodist	[illegible]	[illegible]	[illegible]	[illegible]	[illegible]	[illegible]	[illegible]	[illegible]	[illegible]	[illegible]
Methodist Protestant	[illegible]	[illegible]	[illegible]	[illegible]	[illegible]	[illegible]	[illegible]	[illegible]	[illegible]	[illegible]
Seventh Day Adventists	[illegible]	[illegible]	[illegible]	[illegible]	[illegible]	[illegible]	[illegible]	[illegible]	[illegible]	[illegible]
Presbyterian	[illegible]	[illegible]	[illegible]	[illegible]	[illegible]	[illegible]	[illegible]	[illegible]	[illegible]	[illegible]
Baptist	[illegible]	[illegible]	[illegible]	[illegible]	[illegible]	[illegible]	[illegible]	[illegible]	[illegible]	[illegible]
United Brethren	[illegible]	[illegible]	[illegible]	[illegible]	[illegible]	[illegible]	[illegible]	[illegible]	[illegible]	[illegible]
Lutheran	[illegible]	[illegible]	[illegible]	[illegible]	[illegible]	[illegible]	[illegible]	[illegible]	[illegible]	[illegible]
Congregational	[illegible]	[illegible]	[illegible]	[illegible]	[illegible]	[illegible]	[illegible]	[illegible]	[illegible]	[illegible]
Methodist Episcopal	[illegible]	[illegible]	[illegible]	[illegible]	[illegible]	[illegible]	[illegible]	[illegible]	[illegible]	[illegible]
Total	[illegible]	[illegible]	[illegible]	[illegible]	[illegible]	[illegible]	[illegible]	[illegible]	[illegible]	[illegible]

There is a Union church in Walton which is not included in the above table, making the whole number of houses of worship in the county 60.

The following table shows the census of Eaton County by Townships and Cities, since 1840.

Name of Town.	1840.	1850.	1860.	1870.	1880.	1890.	1894.
Bellevue	[illegible]	[illegible]	[illegible]	[illegible]	[illegible]	[illegible]	[illegible]
Benton	[illegible]	[illegible]	[illegible]	[illegible]	[illegible]	[illegible]	[illegible]
Brookfield	[illegible]	[illegible]	[illegible]	[illegible]	[illegible]	[illegible]	[illegible]
Carmel	[illegible]	[illegible]	[illegible]	[illegible]	[illegible]	[illegible]	[illegible]
Chester	[illegible]	[illegible]	[illegible]	[illegible]	[illegible]	[illegible]	[illegible]
Charlotte	[illegible]	[illegible]	[illegible]	[illegible]	[illegible]	[illegible]	[illegible]
Delta	[illegible]	[illegible]	[illegible]	[illegible]	[illegible]	[illegible]	[illegible]
Windsor	[illegible]	[illegible]	[illegible]	[illegible]	[illegible]	[illegible]	[illegible]
Eaton	[illegible]	[illegible]	[illegible]	[illegible]	[illegible]	[illegible]	[illegible]
Eaton Rapids	[illegible]	[illegible]	[illegible]	[illegible]	[illegible]	[illegible]	[illegible]
Eaton Rapids City	[illegible]	[illegible]	[illegible]	[illegible]	[illegible]	[illegible]	[illegible]
Grand Ledge	[illegible]	[illegible]	[illegible]	[illegible]	[illegible]	[illegible]	[illegible]
Hamlin	[illegible]	[illegible]	[illegible]	[illegible]	[illegible]	[illegible]	[illegible]
Kalamo	[illegible]	[illegible]	[illegible]	[illegible]	[illegible]	[illegible]	[illegible]
Oneida	[illegible]	[illegible]	[illegible]	[illegible]	[illegible]	[illegible]	[illegible]
Roxand	[illegible]	[illegible]	[illegible]	[illegible]	[illegible]	[illegible]	[illegible]
Sunfield	[illegible]	[illegible]	[illegible]	[illegible]	[illegible]	[illegible]	[illegible]
Vermontville	[illegible]	[illegible]	[illegible]	[illegible]	[illegible]	[illegible]	[illegible]
Walton	[illegible]	[illegible]	[illegible]	[illegible]	[illegible]	[illegible]	[illegible]

MISCELLANEOUS SKETCHES.

J. VAN OSDALL.

J. Van Osdall was born April 11th, 1839, in Wayne county, Ohio. He was united in marriage to Mrs. Susannah B. Dixon, February 1st, 1865. In the spring of 1860 they moved to Michigan, settling on a farm in Windsor township, where they still live. Mr. Van Osdall discovered, soon after his settlement on the new farm, that an excellent quality of stone lay hidden beneath the soil and decided to quarry some of it as an experiment. In color, the stone is a very light gray with a faint bluish tint. It is free from lime and iron and does not tarnish as quickly as many other kinds of sandstone. About twenty-five or thirty men will be employed this season in quarrying it.

S. HORNER & SONS.

The Eaton Rapids Woolen Mills, of which S. Horner & Sons are the proprietors, are located on the north end Main St., and represents one of Eaton county's varied industries. This plant succeeds the pioneer carding mill of William Gallery, the change from carding exclusively to woolen manufacturing having been effected some years ago. Yarns, flannels, cassimeres and blankets all of excellent quality are manufactured. The mills consist of a long three story frame building 36x90 feet, a dye house, a boiler house, and a one story frame structure 20x40 feet. The mills usually run the year round and employ, when doing full duty, from twenty to thirty hands at good wages. The mills represent an investment of about twenty thousand dollars and bring to Eaton Rapids a gross amount of some seventy-five thousand dollars each year. Messrs. Horner are also proprietors of the Eaton Rapids electric light plant.

MRS. T. W. DANIELS.

Mrs. T. W. Daniels, nee Anna N. Shord, was born in Alleghany county, New York, in 1841. She was the daughter of John and Catherine Shord who came to Michigan in '44 and settled in the town of Onondaga, Ingham county, where she grew to womanhood, coming to Eaton Rapids to live when but seventeen years of age. She had the good fortune to meet and marry T. W. Daniels who was known in Eaton county as one of the brightest of merchants and business men. Mr. Daniels died Sept. 7, 1891, and by his will showed his great love and the confidence he reposed in his wife by giving her his entire fortune.

WM. B. VAN ALLEN.

William B. Van Allen, who was one of Hamlin township's historic landmarks, was born in Cayuga county, New York, January 10th, 1816. He was the third son of Daniel Van Allen of New York, and was given a very good education in that state. At the age of eighteen he came to Michigan and located in Hamlin township, then called Tyler. Here in a vast wilderness he began the subjugation of the soil and soon made for himself a valuable and comfortable home. He lived the life of an active, enterprising farmer, and was instrumental in the introduction of numerous local improvements. With democratic interests of the state and county, he was actively identified from the beginning. On the 7th of July, 1887, he passed peacefully away at the residence of his son, D. D. Van Allen, mourned by hosts of friends living in Eaton and other counties and states. He was the father of six children, four of whom are deceased. Mrs. Albert Clegg and D. D. Van Allen are of the county's substantial citizens and are still residents of the township in which they were born.

T. M. BISSELL.

The T. M. Bissell Plow Co., No. 115 Canal street, was established in 1848 by James Gallery. In 1882 the name of the plant, owing to the decease of its founder, was changed to James Gallery's Son's foundry and machine shop. A stock company was formed in 1893, for the manufacture of plows, and the new institution received corporated privileges from the Secretary of State in April of the same year. The capital stock is $25,000. The plant gives employment, at good wages, to twenty-five men, and the output, consisting of thirty-eight different styles of plows, is about six thousand a year. The Bissell Plow Co's. principal market is the Eastern and Central states.

LYMAN BENTLY.

Lyman Bentley was born in Gustavus, Trumbul Co., Ohio, December 14th, 1838. His father was a cheese manufacturer of Gustavus, and the family, of whom Lyman was the oldest, consisted of four children. Lyman attended the district schools of Ohio, but at the early age of fifteen began life for himself as a maker of cheese boxes. Later he entered a general store in Wayne township, Ohio, where he spent several winters as a clerk. At the age of twenty-one he secured employment in a dry goods store in Warren, Ohio, where he remained as clerk most of the time for about three years. In 1863 Mr. Bently and his father formed a partnership and purchased a good hotel equipment in Warren and for several years did a successful business. From Warren Mr. Bently went to Louisiana where he spent four years as a farmer. He has resided in Eaton Rapids since 1872, and has given most of his time to the boot and shoe trade. There are many other enterprises however, in which he takes a lively interest. Numerous municipal positions such as city treasurer, president of council, and chief of fire department have been bestowed upon him. He was honored with the presidency of the State Firemen's Association and holds also a seat of membership in the National Association. Mr. Bently was united in marriage to Miss Carrie Decker in March, 1866, at Geneva Falls, Ohio. Mrs. Bently is a cousin of the world renowned William Cody.

MRS. T. D. WILLIAMS.

Mrs. T. D. Williams, of Duttonville, is the widow of T. D. Williams, one of the pioneer physicians of Brookfield. She has, for a number of years, kept the store located opposite the postoffice. Her store is always well stocked with groceries, hardware, boots and shoes, dry goods and necessaries of all kinds. Mrs. Williams has the confidence of her many customers who speak of her as strictly honest and fair in all her dealings with them. She is an earnest worker in the church at Duttonville and gives material financial aid. She is popular and successful as a business woman, esteemed by the community in which she lives for her many excellent qualities.

HORACE B. PERRY.

Horace D. Perry, of Brookfield, was born August 28th, 1825, at Murray, Orleans Co., N. Y. He is a descendant of Ebenezer Perry, who, with three brothers, came from England about 1725. One of the brothers settled in a southern state, one in Massachusetts, and one in Rhode Island. Two grandsons of the latter, Oliver Hazard Perry and Matthew Calbraith Perry, have placed their names high in the history of our country. Ebenezer, the great grandfather of our subject, settled in Connecticut and there married Miss Mary Williams about 1765. To them were born Nathaniel, William, Ebenezer, Asaph, James, Fannie and Mary. The youngest son, James, married Miss Fannie Avery, of Vermont, about 1810. They had six sons, James Atkinson, Oliver Williams, Walter Avery, Harrison G., Horace B., and George, and four daughters, Fannie, Malinda, Sally Amanda, Anna A., and Mary. Horace B. settled in Concord, Michigan, in 1846 and two years later married Miss Leakema Hicks, eldest daughter of Samuel and Betsy (Reynolds) Hicks of Marshall. Of the five children born to them three survive, Cornelia Ann, George Avery and Nathaniel James. The latter has been a school teacher and Inspector of for Brookfield. The former married Dr. W. E. Vanaude now of Sunfield. His mother died when Horace was about eight years old, then he was thrown upon his own resources. He refused to be bound out, and by dint of persevering industry he gained a good education for those days. Mrs. Perry, who was born at Newsted, Erie Co., N. Y., January 12th, 1829, was educated at Rockford, Ill., her earlier home.

Mr. Perry has been honored many times with positions of trust, among them being that of supervisor of Brookfield, which office he resigned in his fifth term on account of poor health. His official as well as his private life will bear closest scrutiny. The home of Mr. and Mrs. Perry has always been a refuge for the unfortunate ones. The hungry never pass from their door unfed or unassisted. Charles, a little foster son, finds a substantial home with them. He attends the district school, studies music and is an affectionate and obedient boy. These esteemed pioneers of Brookfield have lived quiet, unassuming lives, and their strict honesty and intelligent industry make them worthy of the good name which is theirs. The world is made better by the influence of such people as Mr. and Mrs. Horace B. Perry.

GEORGE D. PRAY.

George D. Pray was born in the township of Superior, Washtenaw Co., Mich., February 2, 1848. His parents, Nathan H. Pray and wife, settled in Windsor township in 1837. There were only two families living in the township at this time, Mr. Pray being one of them; the other family had come in some time earlier. In 1842 Mr. Pray moved to Washtenaw county, but after a brief stay of three years returned again to Windsor. Mr. Pray's well equipped farm of two hundred acres is a part of the homestead on which he has lived since between two and three years of age. On Friday afternoon, January 18, 1895, he was attacked by his bull which he was leading, and was so badly injured that he died the following evening. Because of his unselfishness, straight forward honesty, his high purpose in life, and pure character, he was one of the best known and most highly respected men in Eaton county. He leaves a wife, two daughters, a son and an adopted daughter.

JACOB UPRIGHT.

Jacob Upright is a resident of section twenty-one, township of Benton. His birth occurred in Oil Spring, Maulbron Co., Wertenberg, Germany, November 2, 1862. He was educated in the district schools of his native country. In the spring of 1851, John Upright, the father of Jacob, accompanied by his family, emigrated to Oneida, then to Benton township, where he bought a forty acre farm which is now a part of Jacob Upright's possessions and the site of his present home. Our subject remained at home until the first call for volunteers was issued, when, although foreign born, he determined to enter the service of his new country. He joined the 66th Illinois Regiment, Sharp Shooters, and stayed until the close of the war. He is one of the very few men who escaped all the showers of shot and shell, and came out as good a man as when he entered. In May, 1866, he was united in marriage to Miss Sarah Brunn, also of German extraction, and a native of Lewis Co., N. Y. Five children are the fruit of this happy marriage, Eva, Clarence, Ray, Estella and Maud; the first men-

Res. of James Baughman, Charlotte.

Horner's Woolen Mill, Eaton Rapids

S. T. Green, Implements, Charlotte.

Old Maid's Block, Charlotte.

Res. of D. D. Van Allen, Hamlin Twp. Portraits and Old Residence of Mr. and Mrs. Wm. B. Van Allen.

School Bldg., Grand Ledge, 1st Ward.

Sunfield Elevator, Sunfield.

Michigan State Bank, Eaton Rapids

Res. of Rev. W. S. Williams, Charlotte.

District No. 8, Walton Twp.

School Bldg., Grand Ledge, 2nd Ward.

Eaton Rapids City Schools.

Res. of Seth Ketcham, Charlotte.

STONE QUARRY OF J. VANASDALL, DIMONDALE.

tioned is now Mrs. Albert Towe and resides in Charlotte, the other children still residing at home. Mr. Upright's home, as will be seen in the above illustration, is one of Eaton county's finest, and his farm, a well stocked, well equipped tract of two hundred and forty acres, is rich and productive.

J. MIKESELL & COMPANY

Is the title under which the preserving factory of Charlotte is conducted. This institution is situated at the north end of Oliver street, opposite the Michigan Central depot. The plant consists of three large buildings, the first of which is a two story structure, 35x70 feet. The west end of the first floor is equipped as an office, and the packing, shipping and storage rooms, in their respective appartments, occupies the rest of the building. The second is a one story building and measures 35x85 feet, and contains the cooking department, the heavy machinery and the engine house. The third building is a store room for uncanned stock, measuring in size 20x50 feet.

The plant is modern in every detail; the large boilers are of 70 horse-power and are used for cooking, in their season, berries, tomatoes, apples, peaches and green peas. To give the reader an intelligent idea of the extent and usefulness of their new institution, we will say that its daily capacity to put out its products are as follows: Apples, 5,000 gallons; tomatoes, 18,000 cans; and berries, 8,000 to 10,000 cans, and its market is the unlimited territory of the United States.

The force employed is eighty hands, and this number is found inadequate for the business, and the pay-roll per week is $400, and the weekly purchase of stock is at present something like 6,000 bushels of apples at a cost of $1,500. Hereafter the company expect to double its force of hands, likewise its pay-roll and its out put of goods.

The management of the cannery is in the hands of Lundy F. Mikesell, the junior member of the firm, a young and enterprising man, son of J. Mikesell and a native of Charlotte. The book-keeping of the institution, which will be readily seen is no small task, is in the hands of Miss Lula Mikesell, daughter of the senior member of the firm.

Jerrie Mikesell is a native of Ashland, Ohio, where he was born January 28th, 1838. His father, Jos. Mikesell, a native of Pennsylvania, was of German decent, a brick and stone mason in his younger years but a farmer in later life. When our subject was fifteen years old his father emigrated to the present site of Charlotte, Mich. where he lived to the age of ninety years, his demise occurring August 23, 1892. Jerry has lived an active business life, as a grocer, as a farmer, as a real estate dealer and as a manufacturer. His friends call him broad minded, stirring and enterprising and we do him an injustice to omit the statement that his public spirit has has been a leading factor in putting Eaton county at the head of the list in Michigan.

G. H. FOWLER & CO.

This firm consists of G. H. Fowler and wife, photographers and portrait painters, No. 123 Cochrane avenue.

Mr. Fowler was born in New York forty-five years ago. He is the son of Henry Fowler, a farmer of Ontario county. In 1861 he came to Michigan in company with his parents. After several removals they settled in Charlotte in 1874. Mr. Fowler is a selfmade photographer, but he studied portrait painting with leading artists of Michigan and it is by this latest mentioned branch of his business that his reputation first became known. He has been proprietor of his present well known gallery since 1889. The superior quality of his work has earned him a splendid reputation as an artist, which was acknowledged by the National Photographers' Association at the World's Fair convention by the awarding to him of two silver medals. He received another medal in '91 at the National Photographers' Convention held at St. Louis, and is now President of Michigan Photographers' Association which he helped to organized in January this year.

Mrs. Fowler is a practical and experienced operator of the camera and has been identified with the business for eleven years. She is also one of Charlotte's energetic women, and is interested in all enterprises that tend to the best interests of the people. She is a member of the School Board of Charlotte and also county secretary of the State Sabbath School Association. We will add that a large number of the illustrations contained in this work, are engraved from photographs made by Mr. and Mrs. Fowler.

SPENCER C. PHARES.

Spencer C. Phares is the proprietor of the feed barn between Lovell and Lawrence on Bostwick avenue. Every one in the city, and many horse owners in the adjoining neighborhood know where the skating rink was located. Mr. Phares has had this building for two years, and has the finest accommodations for man and beast. He is an industrious gentleman, and treats all his customers with courtesy. He is a native of Cincinnati, Ohio, and served in the Union army for nearly five years. He belongs to the G. A. R., and is deserving of patronage. An event in his military career was the part he took in the capture of Jefferson Davis at Irwinville, Georgia, May 10th, 1865, and two hundred and ninety-three dollars of the one hundred thousand paid by the President for the capture of Davis, was received by Mr. Phares as his share. The records of Congress, shown the writer, substantiate his claim.

THOMAS JENSEN.

Thomas Jensen, feather renovator, has lived in Charlotte for nearly two years and is regarded as absolutely honest in his dealings with his customers. He is an expert as a feather renovator, and no one in Charlotte would hesitate to entrust him with pillows or feather beds. All his customers say he gives perfect satisfaction. He is also the proprietor with Mr. Wilson of a second-hand store. All kinds of new and second-hand goods are bought and sold, of which the firm keeps a great variety constantly in stock. The people of the county are invited to call and see his three hundred dollar renovating machine. Mr. Jensen employs reliable agents only.

CHARLOTTE GREEN HOUSE.

Mrs. Emma J. Church, proprietress, corner Cochrane avenue and Henry street. Four large buildings are required to accomodate this institution. Hot beds almost without number, using over 4,000 square feet of glass to cover them, nearly three-quarters of a mile of piping is used for heating purposes, and in cold weather over twenty-five bushels of coke are burned each day. Mrs. Church is a practical business woman, with a natural aptitude for the growing of flowers and vegetable plants. She has just filled a single order for 400,000 tomato plants. Attention is immediately given to orders of cut flowers or art work for parties, weddings, funerals, etc., her resources enabling her to supply any demand. Nor does she depend solely upon a home market, as she almost daily ships orders to the eastern and western cities. The accompaning illustration of the interior of her pleasant home, shows a beautiful floral display of her own handiwork.

HON. GEORGE N. POTTER.

In Eaton county it would be difficult to find a man so stalwart, so capable of great ideas, so successful in bringing forth results as Geo. N. Potter of Potterville, Charlotte and Lansing. He is a manufacturer, merchant, and farmer and is more successful in each branch of his business than the average man who has his attention confined to but one of the occupations.

Mr. Potter was born in Cayuga Co., N. Y., but has resided in Michigan since three years of age. His early life was beset with all the severe trials that come to poverty-stricken pioneers. In 1844 or '45 his father came to Eaton county and built a log shanty with a roof of troughs and a puncheon floor, neither nails nor boards having been used. The father died in 1846 and George N., then eighteen years old, supported and protected his widowed mother and orphan brothers and sisters. His school career consisted of one three months' term in Vermontville, he chopping wood to pay for his board. He earned fifteen dollars, and received thirty-five dollars more from his mother, and with this amount purchased his first farm, a forty acre tract of government land. Upon this he built a log house, and on March 1, 1849, was married to Miss Martha L. Gladding, a native of New York. He has been sheriff of Eaton county four years, was deputy provost-marshal during the war and has served his county in the state legislature. He was the first to introduce the circular sawmill in the county, was one of the prime movers in securing the Grand Trunk and Michigan Central railroads, the first of which he was for years a director. He is at present the owner of nearly a thousand acres of Eaton county land, is interested in a large factory in Lansing, and also the Benton Manufacturing Co. of Charlotte. He owns a large brick and tile factory at Potterville, a creamery, a flouring mill, a hotel, a brick block and a sawmill in Delta township. His son, John C. of Charlotte, is interested with him in many of his enterprises.

Mrs. Martha L. Potter having died in 1869 he, in 1870, was united in marriage to Miss Mary A. Page of St. Lawrence county, New York.

JACOB MICHEL.

Cigar manufacturer, is of German nativity, his birth occurring in Herdingsfelds, Bavaria, Nov. 19, 1857. March 13, 1882, Mr. Michel came to America and found employment with the large cigar firm of Shotwell & Co. in New York. He soon left this place and in a few months came west, arriving in Charlotte January 1st, 1894. Here he established a factory of his own and it ranks the largest of the four similar institutions located in this city.

Mr. Michel's factory employs four men constantly and turns out an average of three thousand cigars per week, of which about one-third are ten cent goods. His brands are among the most popular and are known as Star Unions, Good 'Nough, Honesty, Emblem, Large Hiawatha, Small Hiawatha, and Charlotte Pride. The first four are five cent goods, long filler, hand made mixed with Havana with Sumatra wrappers. The last three are his popular ten cent brands. Large Hiawatha is a clear Havana filler with Sumatra wrapper; it is a five inch cigar and weighs about eighteen and a half pounds per thousand. Small Hiawatha and Charlotte Pride are four and three-eighths inch cigars with other qualities about the same as the Large Hiawatha. They are exceedingly popular and no flavoring is used in any of them.

W. B. OTTO.

This well known farmer was born Jan. 16, 1844, in Wood county, Ohio. The parents of Mr. Otto were Henry and ——— -Bryan Otto, the former a native of Pennsylvania and the latter of Ohio. William was given a common school education. His father died when he was quite young but he continued to work the farm as the loyal support of his widowed mother until the second call for volunteers was issued from Washington. To this Mr. Otto responded. He was then barely sixteen when he entered the army in which he remained until the close in 1865. At Knoxville he was taken prisoner and confined in Libby prison for several weeks. An exchange of prisoners caused his release and he returned to his regiment, the 111 Ohio Infantry. When he returned from the war in 1865 he located in Eaton county, Mich., and engaged in lumbering and farming. In 1878 on the 16th day of January, Mr. Otto was united in marriage to Miss Celia M. Potter, daughter of the Hon. G. N. Potter, a prominent pioneer citizen of the county. The next year after his marriage he purchased a beautiful farm of 240 acres, known as the G. N. Potter homestead.

American System of Rectangular Survey.

The struggle for independence of the thirteen American colonies with Great Britain, although a successful one, left the colonies with a heavy burden of debt to pay. The fact, however, that several of the colonies (now states) had an interest in what was then known of the Northwest Territory, proved one of the most powerful influences which kept the new born nation from dropping to pieces, and a fruitful means to assist in clearing off the burden of debt.

The four states, Massachusetts, Connecticut, New York and Virginia which claimed all the land north of the Ohio river, west to the Mississippi, agreed (from 1780-1786) to give it to the United States, to be disposed of for the common good, and in 1787 Congress passed an ordinance for the government of this territory, and also for establishing a definite method for the survey and sale of these lands, which were then designated as "Public Lands" to be placed on the market for sale, the proceeds of which were to be principally applied to the payment of the war debt of the Revolution.

To draw up a definite plan for the survey of these "Public Lands" in the Northwest Territory was a difficult problem, as the methods of survey in the different states differed somewhat. Virginia had her regular plats known as "Tomahawk Surveys." Connecticut had a more uniform plan which she had adopted in her survey of the "Western Reserve" in Ohio, part of the territory to which she laid claim. And now as all these different states had ceded all this territory to the general government for the good of all, it became highly necessary that some general and definite method of survey be adopted.

The plan arranged by James Mansfield, surveyor general of the Northwest Territory, was adopted by Congress in 1802. It is so simple and practical that it has received very few modifications by any of the land commissioners since. After the adoption of a definite method of survey the government proceeded to have tracts of this territory surveyed off as the demands of the public required; the first tract surveyed being nearly all in the state of Ohio, the second in Indiana.

The United States Rectangular Survey may be briefly stated as follows:

First, a north and south line is run through the tract determined upon to be surveyed. This line begins at some prominent or easily distinguished point, and is designated as a "Principal Meridian." Then a line running east and west, at right angles with the first line, is run through the tract, called the "Base Line."

The Principal Meridian of our state begins at a point forty-eight miles west of Lake Erie, on a line between Michigan and Ohio, and from there extends north to the City of Sault de Ste. Marie. The Base Line extends from Lake St. Clair to Lake Michigan, forming the south boundary of Eaton County. These lines are run with a "Solar Compass," avoiding the errors of a magnetic needle.

Lines are then run north and south parallel to the Principal Meridian and six miles apart, which divide the territory into long north and south strips called Ranges, which are numbered in their order 1, 2, etc., East of the Meridian, also the same west of it. In Michigan there are 17 Ranges East and 42 West. Across these are run lines six miles apart, parallel to the Base Line, cutting the territory into long east and west strips called Towns, and these are numbered North and South from the Base Line. In Michigan there are 47 Towns north and 5 south.

By this "crosslining" the territory is divided into squares, six miles on a side. Each of these squares is a Congressional Township. Such "Townships" sometimes, but often do not, correspond to the Civil Townships which are known by popular names. The only designation of Congressional Townships is their Range and Town numbers. The system is illustrated by the following diagram:

X is Township 2 North, Range 3 East.
Y is Township 2 South, Range 2 West.
Z is Township 4 North, Range 1 East.

In practice the surveyors did not run the Range and Town Lines their whole length, continuously. The magnetic needle points east of north in Michigan and its variation from north continually changes. Running a line through primeval forests is beset with difficulties. No measurements of such great length can be made exactly. Hence the surveyors began on the Base Line six miles east of the Initial Point, ran a Range Line six miles north as nearly as they could, and then ran a "random line" west to the Principal Meridian, to check their work. Then they ran back to their Range line, marking section and quarter-section corners as they went, and so proceeded to lay out the next township north; and so on east and west of the Principal Meridian.

But as they ran north, on account of the fact that all lines running north continually approach each other and come together at the pole, every township was narrower at the north than at the south. To prevent this error growing, every fourth Town Line north and every fifth Town Line south of the Base line is called a "Correction Line," and on these a fresh start is taken with distances full six miles east and west. "Auxiliary Meridians" were also established at every eighth Range Line.

After the tract is thus surveyed into townships six miles square, the townships are divided into thirty-six tracts, called "Sections," each containing one square mile, more or less.

The sections were run off very much as were the townships, using each township's east Range Line and south Town Line as bases. Commencing one mile west of the southeast corner of the township, the surveyor runs north a mile, then east a mile to the east range line and corrects back to the northwest corner of the section. He sets a quarter post (or a half mile post) on the west line of the section at forty chains north of the starting point, and sets the quarter post on the north line of each section, half way between the northwest and northeast section corners. The surveyor proceeds to run off the remaining sections on the east tier, up to the north line of the township, placing the last section corner where his north and south line intersects that north town line, whether this point is east or west of the section corner previously established in the township survey. The distance between the two corners, if any, is called the "jog," and is recorded.

In the more recent government surveys in the west, there is no "jog" left, the surveyor being required to close his lines at the section corners on the north and west lines of each township. The section surveyor establishes no quarter or half mile posts on the north line of any of the sections on the north and west sides of the township. Each tier of sections in the township is run off in this manner, except the last two, which are run off together.

On account of the errors previously mentioned no township will divide into thirty-six exact sections, and in the sectional survey new errors arise. These errors are all run into the north and west tiers of sections, which are called "Exterior" or "Fractional Sections," because they contain the excess or deficiency of land in the township, and this apparent excess or deficiency is always thrown into the last quarter mile, lying next to the township lines on the north and west. The other sections are called "Interior" sections, and are intended to be full six hundred and forty acres each, but they nearly always exceed or fall short of this amount.

The government sub-divisions of the section (although they are not actually surveyed by the government surveyor) by which the lands are sold, are "quarter" sections, or one hundred and sixty acres; "half-quarter" sections, or eighty acres, and "quarter-quarter" sections, or forty acres. The section is divided into quarters by running a straight line north and south and one east and west between the quarter posts on the sides of the section. The quarter sections are "halved" by running a straight line north and south or east and west (whichever way is wished to divide it) from points midway by measurement of opposite sides. The quarter sections are quartered by running lines north and south and east and west between points at the center of each side of the quarter section. Other smaller sub-divisions can be made on the same principle.

It will be seen from this that if a section is perfectly square and contains the exact number of acres, that this method would subdivide it into tracts of equal areas, but it hardly ever occurs that a section is exactly square or contains the exact number of acres. Consequently, it almost always occurs that the sub-divisions will differ more or less in quantity. But the government has established this as the only method by which the sub-divisions shall be made, making the eight corners established on the exterior lines of each section "the corners," however inaccurate they may be.

In order that no one purchasing lands from the government may suffer injustice in expecting to get the actual number of acres intended to be in each sub-division, the government sells all of its lands on the condition that each one of these sub-divisions contains so many acres, "be the same more or less," according to the government survey. And this rule follows the future transfer of the lands, where they are sold and described in "Government Descriptions," whether the words "more or less" are mentioned in the deed of conveyance or not. The method of description under this system is exact, and simple when once understood. The township is described, as previously stated, by the numbers of its town and range. The sections are numbered from one in the northeast corner to six in the northwest corner, then the next row below that from left to right, and so on back and forth to thirty-six, in the southeast corner.

The sub-divisions of the section in the following diagram as it is divided into "Government Descriptions," are each described in brief on the diagram. The one marked X we will describe in full as a sample of all.

Section 26, Township 36 N., Range 11 East.

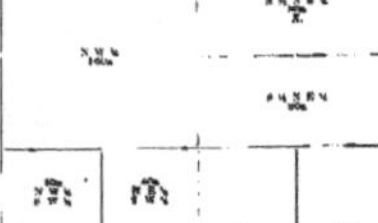

"The north half of the northeast quarter of section 26 township 36 north, range 11 east of the Principal Meridian."

The ne ¼ and nw ¼ of each section lying on the north side of the township are described as the fractional ne ¼ or the fractional nw ¼, and the sw ¼ and the nw ¼ of each section lying on the west side of the township are described as fractional nw ¼ or fractional sw ¼ of such sections.

If any of the fractional quarters on the north side of a township are divided into halves by an east and west line, the south half is made eighty rods wide and the north half takes the excess or deficiency and is described as the fractional n ½. If they are divided by a line running north and south each half is described as either the e ½ or the w ½ of the fractional ne ¼ or nw ¼.

Of the fractional quarters on the west side of the township the descriptions would be the reverse, as they are divided by a north and south or an east and west line. When a section contains a lake which was meandered out in the original survey, the fractional pieces in each quarter section were numbered as lots, and sold by the government as lot No. — in — quarter, sec. —, tp. — N., R. — E.

Land may also be described by "Metes and Bounds," that is, the actual beginning of the lines and actual measurements being given. Thus: "A parcel or tract of land lying in the southeast quarter of Sec. 23, Twp. 36 North, Range 5 East, commencing at a point ten chains east of the southwest corner of the southeast quarter of said Sec. 23, running thence east ten chains, thence north twenty chains, thence west ten chains, thence south twenty chains to the place of beginning, containing twenty acres."

A tract running the whole length of any side of a square or rectangular piece of land, as a quarter-section, half-quarter or quarter-quarter, can be definitely described as so many acres off of the E side, or W side, or N side, or S side, whichever side it may be. But if the tract does not run the whole length of a side, that style of description would be wrong.

There is one very common error in the description of land, and that is many notaries public, attorneys and justices of the peace, where there may be an eighty acre tract or any other government subdivisions to be divided among different parties, who are unwilling to have each of their different interests surveyed before their deeds are made, naturally do this in their minds, that if it is a forty acre tract it must be eighty rods square, or if it is eighty acres it is one hundred and sixty rods long and eighty rods wide, or if one hundred and sixty acres that is one hundred and sixty rods square, which in our government sub-divisions hardly ever occur exactly, so in dividing the government subdivisions, as if they were exact in measurement on each side, the different pieces will overlap each other as they are described, or leave a surplus not conveyed to any one of the parties. Real estate should be so conveyed that there could be no question as to its metes and bounds when it is surveyed.

TABLE OF MEASUREMENTS.

LINEAR.

16½ Feet = 1 Rod.
80 Rods = ¼ Mile.
160 Rods = ½ Mile.
320 Rods = 1 Mile.

7.92 Inches = 1 Link.
100 Links }
66 Feet } = 1 Chain.
4 Rods }

80 Chains = 1 Mile.

SQUARE.

272¼ Square Feet = 1 Square Rod.
160 Square Rods = 43,560 Square Feet = 1 Acre.
640 Acres = 1 Square Mile = 1 Section.

1 Square Acre is { 12.65 Rods Square.
{ 208 Feet, 8½ Inches Square.
{ 3 Chains, 16¼ Links Square.

CHARLOTTE.

2 G. H. Spencer
3 S. M. Cove
4 G. H. Fowler
5 Jacob Michel
6 Mrs. Anna Bush
7 W. Geddes
9 D. B. Alinger
10 H. A. Blackman
11 C. E. Chappel
12 E. H. Bailey
13 Jessie Mikesell
14 J. W. Munger
15 C. M. Jennings
16 A. L. Nichols
17 James Bryan
18 F. M. Stevens
19 M. F. Warren
20 D. A. Castelin
21 Mrs. J. F. Terrill
22 John L. Miller
23 Seth Ketcham
24 G. A. Williams
25 Geo. L. Sing
26 Geo. Bush
27 Frank G. Smith
28 D. L. Freeman
29 L. P. Bissell
30 Jas. Gillingham
31 B. J. Culbertson
32 David Barr
33 Geo. W. Rowley
34 A. B. Betz
35 G. S. Beardsley
36 Mrs. G. S. Beardsley
37 Homer Bryan
38 Horton Bryan
39 J. G. Miller
40 L. O. Smith
41 A. D. Baughman
42 Joseph Lang
43 Myer Vineberg
44 Gideon Cogsdill (Deceased)
45 J. H. Rasey
46 A. B. Allen
47 J. M. Baron
48 A. H. Munson (Deceased)
49 H. F. Reynolds
50 Henry Mull
51 S. C. Hayes
52 Chas. Foster
53 Jas. Gouldsborough
54 J. F. Terrill
55 Frank A. Ells
56 Geo. A. Perry
57 Wm. H. Reynolds
58 D. W. Warren
59 Rev. W. B. Williams
60 J. J. Curtis
61 Albert Murray
62 C. M. J. Young

KALAMO TOWNSHIP.

1 A. B. Swift
2 Mrs. C. H. Wells
3 C. H. Wells
4 Mrs. Thos. Lyon
5 Thomas Lyons
6 Willard Mead
7 W. F. Granger
8 Mrs. A. B. Swift
9 S. A. Petts
10 Mrs. Stephen Benedict
11 J. D. Butler

CARMEL TOWNSHIP.

12 Mrs. Peter Horn
13 Peter Horn
14 Mrs. David A. Grier
15 David A. Grier
16 Chas. B. Peters
17 Jacob Dawson
18 Mrs. Harris Cooper
19 Harris Cooper
20 Mrs. Erastus King
21 Erastus C. King (Deceased)
22 Frank L. King
23 Mrs. S. G. White
24 S. G. White
25 Fred. Schneckenberger
26 Jacob Schneckenberger
27 C. H. Griest
28 P. M. Mason
29 J. Q. Griffith
30 Mrs. W. A. Case
31 W. A. Case
32 Mrs. C. H. Case
33 C. H. Case
34 M. W. Cooper
35 Mrs. M. W. Cooper
36 Mrs. G. H. Wade
37 G. H. Wade
38 Isaac Krusen
39 Mary Krusen

CHARLOTTE (Continued.)

40 Mrs. G. H. Fowler
41 C. S. Jackson
42 Thos. Jensen
43 Floral Piece from Charlotte Greenhouse
44 Miss M. Kinne
45 Miss C. Kinne
46 Belle M. Perry
47 Mrs. C. M. Young
48 D. C. Hoedemaker
49 Wm. Long (Deceased)
50 C. B. Lamb
51 H. H. Curtis
52 B. F. Belding
53 Mrs. Dr. F. A. Weaver (Dec
54 Mrs. Emma J. Church

EATON RAPIDS

1 H. H. Widger
2 Mrs. H. H. Widger
3 H. J. Milbourn
4 A. Osborn
5 Samuel Andrsky
6 John M. Corbin
7 Lyman T. White
8 W. F. Stirling
9 H. P. Webster
10 H. C. Minnie
11 J. C. Shaw
12 W. R. Merritt
13 Chas. Rayner
14 I. N. Reynolds
15 P. C. Birney
16 G. B. Blair
17 L. A. Bentley
18 Arthur Gallery
19 Geo. Minnie
20 Chas. Wack
21 C. F. Fairchild
22 Geo. D. Wilcox
23 Wm. Smith
24 J. D. Filmore
25 Hon. Wm. Miller
26 J. Dowe Trimmer
27 A. V. Roeher
28 Scott H. Rurabeck
29 Wesley Vaughan
30 F. H. De Golia

GRAND LEDGE.

31 W. Ellsworth Davis
32 F. C. Arms
33 F. C. Beach
34 W. J. Babcock
35 G. W. Irish
36 J. M. Burtch
37 Geo. N. Berry
38 Mrs. Vólorous Kent
39 Volorous Kent
40 W. C. Westland
41 A. B. Schumaker
42 E. T. Astley
43 Robt. Astley

WINDSOR AND DIMONDALE

WINDSOR TOWNSHIP.

1 Wm. J. Bateman
2 Mrs. Wm. J. Bateman
3 Mrs. Jas. Rann
4 James Rann
5 Chas. R. Lewis
6 Mrs. Rebecca Tramp
7 Mr. John Tramp
8 O. H. Barber
9 Mrs. O. H. Barber
10 Charles Hull
11 Mrs. Chas. Hull
12 Mrs. F. B. Skinner
13 Frank B. Skinner
14 J. P. Miller
15 Mrs. J. P. Miller
16 T. M. Sloan
17 Mrs. T. M. Sloan
18 C. E. Norton
19 Mrs. C. E. Norton
20 Mrs. James Ure
21 James Ure
22 Mrs. Dr. E. S. Walford
23 J. G. Schmidt
24 Mrs. J. G. Schmidt
25 Geo. W. Rose
26 Mrs. G. H. Shippard
27 G. H. Shippard
28 A. D. Holmes
29 Albert F. Porter
30 Mrs. A. F. Porter
31 Freeman G. Pray
32 Frank J. Spalford
33 Mrs. John Vanzandt
34 John Vanzandt
35 Mrs. A. D. Carlton
36 Hon. A. D. Carlton
37 J. L. McCready
38 Mrs. N. P. Bateman
39 F. E. Phinney
40 Ee'g Pray
41 John Hetrick
42 N. P. Bateman
43 Mrs. Albert Shotwell
44 Wm. Jerkies
45 Mrs. Wm. Jerkies
46 Geo. Pray
47 Mrs. Geo. Pray
48 Mrs. F. G. Pray
49 Mrs. O. D. James
50 O. D. James
51 Mrs. A. W. Strobel
52 A. W. Strobel
53 Silas French
54 Mrs. Silas French
55 Whitman Hull

ONEIDA TOWNSHIP.

1 Mrs. W. H. Sutherland
2 W. H. Sutherland
3 Mrs. Geo. W. Nichols
4 Geo. W. Nichols
5 Chas. V. Fuller
6 Isaac S. Taylor
7 Mrs. John Ewing
8 Daniel Strange
9 Mrs. C. A. Patterson
10 C. A. Patterson
11 Michael Feess
12 Mrs. Michael Feess
13 Wm. Brauger
14 Mrs. Wm. Brauger

ROXAND TOWNSHIP.

15 J. V. O'Neil
16 Mrs. Thomas Vickery
17 Thomas Vickery
18 James Newark
19 Adeline E. Newark
20 Mrs. D. P. Fuller
21 D. P. Fuller
22 Bishop Haddix
23 Mrs. Bishop Haddix
24 Mrs. M. D. Merriam
25 M. D. Merriam
26 Perry Trim
27 Mrs. Perry Trim
28 George Rimmel
29 Mrs. Geo. Rimmel
30 Mrs. L. W. Hoag
31 Rev. L. W. Hoag
32 Mrs. D. V. Helms
33 D. V. Helms

BENTON TOWNSHIP.

34 Ezra Palmiter
35 Geo. P. Huff
36 Geo. S. Cady
37 S. M. Honer
38 Perry F. Hines
39 Mrs. John Woodworth
40 John Woodworth
41 Mrs. C. T. Ford
42 C. T. Ford
43 V. D. Murray
44 Mrs. Augusta Murray
45 Alva P. Claflin
46 Mrs. Alva P. Claflin

DELTA TOWNSHIP.

47 John Berner
48 Wesley Eldred
49 George Lee
50 Evander L. Drake
51 T. Huxtable
52 Mrs. Evander L. Drake
53 Mrs. J. W. Dann
54 Mary A. Dann
55 Lewis J. Dann

VERMONTVILLE AND VERMONTVILLE TOWNSHIP.

1 Eli P. Fashbaugh
2 Mrs. E. P. Fashbaugh
3 W. P. Viele
4 Mrs. Ellen Viele
5 J. N. Hawkins
6 Victor C. D. Hawkins
7 Mrs. E. F. Tubbs
8 E. F. Tubbs
9 Chas. Hull
10 Wm. M. Griswold
11 Mrs. Jonathan E. Lake
12 Jonathan E. Lake
13 Eugene Carey
14 W. C. Alsover
15 Geo. L. Lamb
16 A. R. Williams
17 Homer G. Barber
18 H. S. Dickinson
19 Mrs. H. S. Dickinson

CHESTER TOWNSHIP.

20 Ervam R. Martin
21 Mrs. S. W. Harmon
22 S. W. Harmon
23 Kelley Bosworth
24 John A. Rich
25 Geo. R. Gardiner
26 Mrs. Geo. R. Gardiner
27 Chas. Allen Martin
28 M. F. Young
29 Mrs. M. F. Young

SUNFIELD AND SUNFIELD TOWNSHIP.

30 Daniel Hulett
31 Miss Edith Hulett
32 Mrs. D. Hulett
33 Schuyler Weeks
34 H. B. Sackett
35 Wm. Edwins
36 Albert Hunter
37 J. K. Hunter
38 W. R. Hager
39 David Smith
40 Mrs. Rachel Welch
41 Geo. H. Cheetham
42 Mrs. E. M. Snyder
43 J. Bona Peabody
44 Mrs. J. B. Peabody
45 Geo. V. Hildinger
46 John H. Palmer
47 Henry Chatfield
48 Alden Childs
49 A. B. Bishop
50 L. A. Wilson
51 W. B. Bera
52 L. G. Lemmon
53 Mrs. W. E. Vanande
54 Deloss Bishop
55 Mrs. Deloss Bishop
56 Peter Chatfield

HAMLIN TOWNSHIP.

1 Samuel Hamlin
2 Mrs. D. D. Van Allen
3 D. D. Van Allen
4 David Walter
5 David B. Hamlin
6 Mrs. Eli Walter
7 Eli Walter
8 V. M. Smith
9 N. T. Taylor
10 Mrs. O. B. Lake
11 O. B. Lake

BROOKFIELD TOWNSHIP.

12 Geo. Tulip
13 J. A. Vanauda (Deceased)
14 Mrs. J. A. Vanauda
15 Horace B. Perry
16 Mrs. H. B. Perry
17 Mrs. N. J. Perry
18 N. J. Perry
19 Mrs. Jos. Myers
20 Jos. Myers
21 Mrs. Dr. D. T. Williams
22 G. Fuller
23 Mrs. Ezra D. Spotts
24 Ezra D. Spotts
25 John D. Kay
26 Fridolin Webber
27 Joseph Webber
28 Mrs. Henry Livingston
29 Henry Livingston
30 C. D. Peters

EATON RAPIDS TOWNSHIP.

31 Elton E. Spears
32 Mrs. Cary Ranney
33 Cary Ranney
34 C. W. Stevens
35 Mrs. B. L. Bentley
36 B. L. Bentley
37 Wm. Spicer
38 Mrs. Wm. Spicer
39 Mrs. A. D. Saxton (Deceased)
40 A. D. Saxton
41 Mrs. Levi Rogers
42 Levi Rogers
43 S. C. Mix
44 Guy Parker
45 Mrs. Guy Parker
46 Mrs. C. E. Bennet
47 C. E. Bennet
48 E. B. Spears

EATON TOWNSHIP.

49 Josiah Miller
50 Michael Merkel
51 Anna McFarland
52 Mrs. Josiah Miller
53 Mrs. L. D. Dickenson
54 Benj. Spotts
55 Mrs. Benj. Spotts
56 Mrs. A. Hoffner
57 A. Hoffner
58 Mrs. Abraham Lee
59 Abraham Lee
60 Mrs. Jas. French
61 James French

BELLEVUE TOWNSHIP.

1 P. G. Hemenway
2 Anna Hemingway
3 Guy W. Monroe
4 M. P. Shumway
5 Nicholas Simons
6 Mrs. Nicholas Simons
7 J. A. Spaulding
8 Mrs. J. A. Spaulding
9 Joel Kelley
10 Mrs. Joel Kelley
11 W. West
12 F. E. Andrews
13 L. E. Spafford, Jr.
14 Abram B. Hoyt
15 James Mulvaney
16 Mrs. Horatio Hall
17 Horatio Hall
18 Jacob W. Depuy
19 Geo. P. Stevens
20 Curtis A. Day
21 W. E. Hult
22 Albert J. Sawyer
23 A. E. Fitzgerald
24 J. R. Hall
25 Hitam M. Allen
26 John H. York
27 T. E. Robinson
28 N. H. Johnson
29 Allen Havens

OLIVET.

30 Wm. Farlin
31 Mrs. Wm. Farlin
32 A. W. Walker
33 Edward Waterson
34 Henry Shallar
35 B. W. Pasch
36 Edwin N. Ely
37 Fred N. Ely
38 Prof. J. Estabrook (Deceased)

WALTON TOWNSHIP.

39 Clinton Hockenberry
40 Geo. W. Sweet
41 Dutger Mott
42 Chester Smith (Deceased)
43 P. A. Fisher
44 W. C. Roberts
45 F. Waggoner
46 S. K. May
47 W. Perry Ogden
48 J. M. Dillon
49 Frank Martin
50 Jacob Reasoner
51 J. W. Reynolds
52 O. Osborn
53 Mrs. Susanna Ogden

DIRECTORY

OF THE

RESIDENT FREEHOLDERS

Of Eaton County, Michigan.

CITY OF CHARLOTTE.

Abell, Geo. W., farmer, So. Cochrane
Acker, S. J., V. S. So Sheldon
Abell, E. J., laborer, So. Cochrane
Alnger, D. B., D. A. Gen., W. Harris
Allen, A. B.
Allen, G. B., M. D., So. Cochrane
Allen, Sarah J., M. D., W. Lawrence
Allen, John, blacksmith, N. Cochrane
Allen, Addie, school teacher, N. Oliver
Allen, Henry, farmer, So. Sheldon
Allen, Eliza, widow, So Cochrane
Alton, Henry, ret'd far., So. Sheldon
Ames, L. H., farmer, So Cochrane
Ames, Harriett S., widow, W. Henry
Ames, Jason, feed store, So. Cochrane
Anderson, Mariah, N. Sheldon
Anderson, Wm., teamst'r, So. Sheldon
Adams, Sarah S., widow, N. Sheldon
Arnold, I. B., farmer
Annabel Heirs, So. Sheldon
Barrett, C. S., ex p. m., So Sheldon
Bailey, B. H., bookstore, Bostwick
Baker, Frank O., dep. clerk, N. Oliver
Barkhuff, Cynthia, wid., N. Bostwick
Baughman, Jennie M., wid., N. Clinton
Baughman, J. A., bl'ks'th, So. Sheldon
Baughman, C. K., bl'ks'th, W. Lovett
Baughman, A. D., dry g'ds, N. Clinton
Baker, J. H., ret. far., N. Clinton
Barber, B. H., carpenter, N. Clinton
Barber, Miles, laborer, N. Washington
Barber, A. M., hard ware, E. Lovett
Barber & Spencer, h'dware, Cochrane
Bathyte, Edwin, laborer, N. Clinton
Barney Geo. J., dry goods, W. Lovett
Barringer, Christ, farmer, Forest
Baker, N. H., teamster, N. Oliver
Babcock, B. H., pumps, Pearl
Bason, J. H., carpenter, W. Prospect
Barnes, Sylvia A., widow
Barnum, Mrs. O. H. dr'r, S. Cochrane
Ballard, Josephine, So. Cochrane
Bell, Ira, b'ksmith
Bates, A. W., drummer, W. Henry
Beckman, W. M., supervisor, ins. and real estate, E. Cochrane
Bement, J. B., farmer, W. Henry
Belcher, F. B., capitalist, W. Lawrence
Benton, Geo. E., carpenter, Lansing
Beard, Harrison, farmer, W. Henry
Belcher, Mary C., wid., N. Cochrane
Belger, Elizabeth, wid., N. Cochrane
Bennett, Philip, oil dealer, S. Sheldon
Bennett, Chas., manf'r, W. Seminary
Beals, L. J., laborer, N. Clinton
Beardsley, O. S., phot'gr, S. Cochrane
Beardsley, Mary E., p't'g'r, S. Cochrane
Beals, Horace, ret far., So. Sheldon
Bentley, Susan, weaver, W. Henry
Bellinger, Geo., organs, Pleasant
Belger, James E., drayman, Prairie
Bleach, L. F., Editor Republican
Billingsly, O. Heirs, N. Oliver
Billingsly, Rebecca, So. Clinton
Biggs, Lewis, Jas. at L., N. Bostwick
Bliss, Hiram, shoemaker, So. Clinton
Bishop, Jacob, farmer, So. Sheldon
Blakers, Geo.
Biel, Bert, painter, High
Bingham, A. L., ret farmer, Lincoln
Blackmar, H. A., dr'g'st, W. Lawrence
Black, Alexander, carp'tr, N. Sheldon
Bliss, Ida, dressmaker, Lansing
Bower, J. W., baker, N. Clinton
Bolin, B. H., lumberman, N. Cochrane
Bosworth, W. J., lumb'rman, E. Henry
Bodine, Emma, widow, So. Sheldon
Bowie, Wm., engineer, W. Seminary
Bottomley, F. B., Warren
Bowers, Halsey, painter, E. Lovett
Boyles, Wm., agricultural implements
Brenger, John S., sheem'kr, Pleasant
Brumley, Jennie, So. Cochrane
Bretz, Mary A., widow, N. Cochrane
Brevette, Louisa, widow, Forest
Bretz & Co., mer. tailors, S. Cochrane
Bretz, A. D., grocer, So. Cochrane
Britten William, painter
Brown, Mrs. S. S., widow, Pleasant
Bradley, A. A., horse br'dr, N. Clinton
Brasier, Robert, laborer, Lincoln
Brown Bros., clothiers, So. Cochrane
Baldwin Geo. W.
Brugh, Chas. E., clerk, So. Sheldon
Brugh, Joseph, clerk, S. Seminary
Bryan, Jesse, laborer, E. Henry
Bryan, James, farmer
Bryan, Sarah J., E. Henry
Bryan, T. D., ret. farmer, E. Henry
Bryan Bros., (Homer)(Horton) Charlotte Leader
Buck, B. M., horse br'dr, N. Sheldon
Butler, Orsamel, laborer, Pleasant
Burwick, J. W., carpenter, E. Henry
Bundy, G. F.
Burkhead, Samuel, teacher, E. Henry
Burnett, Catherine, Heirs of, So. Cochrane
Byers, W. J. horse farrier, Washington
Bunting, William
Butler, Chas.
Carpenter, Cyrus, farmer, Pleasant
Callister, John, bl'ksm'th, N. Bostwick
Casterlin, D. A., Peonis, N. Cochrane
Campbell, J. M., laborer, N. Cochrane
Campbell, O. F., S. D. A. preacher, Prairie
Cary, Nancy, nurse, N. Cochrane
Carpenter, B. A.
Carpenter, Mary L.
Carpenter, F. T., N. Forest
Carey, Thomas, pianist, So. Sheldon
Carey, G. H.
Casler, Frank, laborer, W. Seminary
Casher, Susanna, W. Seminary
Casher, Mrs. Dan, W. Seminary
Church, E. T., President Merchants National Bank, So. Cochrane
Chappell, Osman, money loaner, So. Sheldon
Chappell, C. E., brick and tile, So. Sheldon
Cherry, Joseph
Childs, J. M., barber
Childs, W. W., farmer, N. Cochrane
Chase, D. S., millright, Lansing
Chevalier, Harry W., laborer, Lansing
Claver, Jacob B., teamster, S. Foote
Claver, Elizabeth, N. Sheldon
Clark, M. E., coal dealer, N. Sheldon
Clark, Phebe H., So. Clinton
Clark, Frank E., music tchr, S. Clinton
Clase, A. P., laborer, N. Clinton
Clark, Alice
Clay, Lizzie J., bdg house, N. Cochrane
Cleveland, L. C., laborer
Core, Samuel M., lumber, N. Bostwick
Cover, Anna, farmer, N. Washington
Conrad, J. H., laborer, Lansing
Corbin, Laura A., widow, E. Henry
Coy, E. L., clerk, W. Henry
Collins, John F., drugs, N. Clinton
Collins, Mrs. Jennie, hswfe, Foote
Collins, A. B., drugs, Pleasant
Collins, G. V., drugs, W. Lovett
Collins, James
Collins, C. V., laborer, Pleasant
Collist, J. L., hardware, So Cochrane
Cogswill, C. H., farmer, N. Cochrane
Cogswill, Mrs. G., N. Clinton
Cook, Mary E., dr'ssm'kr, So. Sheldon
Cook, C. M., capitalist, W. Lawrence
Cook, Jas. C.
Conant, W. H., hotel, N. Oliver
Conklin, Chas., laborer, N. Clinton
Cobb, Frank, lumber, Pleasant
Cole, J. D., farmer, So. Oliver
Cooley, Levi M., mason, E. Prospect
Corell, A. O., pat. fence, So. Sheldon
Corell, H. A., laborer, Lincoln
Cornell, E. S.
Cornell, Geo. P., Seminary
Cartwright, J. W., sawyer, N. Sheldon
Crosier, Chas., shoem'ker, So. Clinton
Crosier, S. P., heirs of, So. Clinton
Crant, John, N. Cochrane
Culbertson, B. J., ins. agt., Pleasant
Culp, Wm. A., laborer, N. Sheldon
Culp, A. J., sawyer, Maple
Curtis, J. T., laborer, W. Seminary
Curtis, J. J., manf'r, Pearl
Curtis, H. H., manf'r, Warren av.
Cross, H. W., laborer
Curtis, L. O., dressmaker, Pleasant
Curtis, Clara, dressmaker, Pleasant
Curry, Samuel, farmer, Prairie
Curtis, E. F., laborer, E. Henry
Cunningham, William
Dunning, F. W.
Dale, Elizabeth, widow, N. Oliver
Davidson, Mrs. Levi, Robinson av. w.
Dacous, J. H., carpenter, W. Prospect
Davis, Fred, laborer, Maple
Davis, Van C., moulder, N. Cochrane
Davis, C. W., carpenter, Prairie
Dart, William, laborer, N. Clinton
Daron, J. M., boots & shoes, W. Lovett
Depew, Lemuel, farmer, Lovett
Dennis, Emmet, butcher, N. Sheldon
Dean, F. A., atty; con. at Naples, N. Sheldon
Dean, Jonathan, ret. far., N. Sheldon
Dean, W. B., hotel, So. Cochrane
Dell, Geo. S., laborer, John
Dexter, E. L., foreman at Bennett's furniture factory, Lincoln
DeFoe, Emma, h'sewife, W. Seminary
Dendleon, Priscilla, wid., So. Sheldon
Dickinson, M. L., carp'tr, N. Wash.
Dick, I. C., farmer, Lansing
Donavon, D. J., salt'mak'r, N. Oliver
Dolph, Mary, housewife, Forest
Dolson, J. L., carr'ge fact., E. Henry
Dolson, D. B., carr'ge fact., Pleasant
Dolson, W. E., carr'ge fact., Pleasant
Dolson, J. L., Jr., carr'ge, W. Lawrence
Dobovan & Packard, furniture, So. Cochrane
Dudley, W. H., ret. b'ws. mer., N. Cochrane
Durand, Marietta, wid., N. Cochrane
Duffy, James, laborer, So. Clinton
Dunlap, E., well pumps, Pleasant
Dunlap, W., ret. har'nm'kr, N. Oliver
Eaton, M. D., drayman, Prairie
Eastman, J. W., ret. real est., Prairie
Eddy, J. F., wag'mkr, W. Lawrence
Egbert, James, mason, W. Seminary
Eile, F. A., postmaster, E. Lovett
Ells, Mary L., housewife, W. Henry
Ells, Geo. W., laborer, W. Henry
Emery, H. J., M. D., N. Sheldon
Emery, Fred, druggist, N. Cochrane
Falgo, Samantha, widow, N. Oliver
Falkner, W. P., ret. far., N. Cochrane
Peal, Lucinda, housekeeper, E. Henry
Ferry, Hattie
Feldbaw, L. W., laborer, W. Lawrence
Ferguson, S., minister and farmer, W. Lawrence
Fisher, Julia, dressmaker, Cochrane
Finch, Mary, widow, N. Bostwick
Fifield, Lyle, teamster, McClure
File, William, laborer, Austin
File, Fred J., shoemaker, So. Oliver
File, Mrs. John, widow, Prairie
Finch, William, retired, Lansing
Fisher, Solomon, laborer, W. Seminary
Fisher, C. M., widow, Cherry
Follett, G. J., farmer, So. Sheldon
Foote, Geo. W., druggist, So. Sheldon
Foote, E. A., attorney, N. Washington
Foote, Francis A., ho'sewife, N. Wash.
Fox, Garry C., attorney, E. Lovett
Ford, Albert, R. R. Agent, Mayor, N. Bostwick
Ford, Edwin, laborer, W. Seminary
Forbes, Maggie, widow, W. Lawrence
Foreman, Geo., ret. hardware merchant, So. Cochrane
Foreman, Grue, housewife, So. Cochrane
Forney, D. V., laborer, N. Sheldon
Foster, Mrs. Wm. T., wid., W. Warren
Follett, Jacob
Foster, Chas., dry g'ds, So. Cochrane
Foster, Augusta, dressmkr, N. Sheldon
Foster, N. B., milliner, E. Lawrence
Foutz, Mrs. J. T., wid., N. Cochrane
Fox, Mathias, carpenter, W. Henry
Fox, Cynthia, widow, E. Seminary
Fowler, G. H., photographer, Prairie
Fowler, Lyman, barber, N. Sheldon
Freeman, L. R., fruits and fowls So. Cochrane
Frase, F. B., carpenter, So. Oliver
Frank, John, carpenter, So. Sheldon
Friesner, Henry, restaurater & ex-saloonist, So. Sheldon
Frost, A. G., carpenter
Fuller, Mrs. W. T., wid., N. Cochrane
Fuller, W. B., laborer, Lincoln
Fullerton, J. C., carpenter, E. Lovett
Fullerton, Sarah J., h'sewife, E. Lovett
Fuller, W. D., night police, Lincoln
Gavit, S. W., police, Lansing
Gamble, Frances, housewife, W. Seminary
Gamble, James, laborer, W. Seminary
Gamble, James, ret far., W. Foote
Gale, Fidelia D., Mrs. widow, Pleasant
Gardner, Lewis, 2d hand store, W. Lawrence
Gardner, Geo. F., W. Lovett
Garber, I. P., peddler, Pearl
Garber, Samuel, ret. farmer, Pearl
Gatling, R. A., register deeds
Gage, M. H., druggist, So. Cochrane
Garrison, Mrs. L., wid., W. Seminary
Gillingham, James, blacksmith, So. Sheldon
Gebbles, W., mer. tailor, N. Cochrane
Gibbons, Chas. F., foreman furniture factory, W. Foote
Gibbons, J. H., travelling salesman, S. Bostwick
Gilchrist, D. T., sec. foreman, Pearl
Gilbert, N. E., shoemaker, So. Oliver
Gordon, Sarah S., wid., W. Lawrence
Goodby, F. M., dry g'ds, N. Cochrane
Goodrich, D. D., sawyer, Pearl
Goodrich, A. J., small fruits, W. Seminary
Goodspeed Bros., boots and shoes, So. Cochrane
Gouldsborough, James, ret. farmer, So. Sheldon
Granger, Joseph, mason, N. Sheldon
Granger, L. C., peddler, N. Henry
Gregory, Hortense, housewife, Prairie
Greenman, Katie, dr'smkr, Pleasant
Greenman, James, sto. deal. Pleasant
Green, Mary E., M. D., W. Lawrence
Green, Alonzo, farmer, W. Lawrence
Green, Ann, widow, W. Lawrence
Green, Lydia, widow, Prairie
Green Homer, telegrapher, E. Lovett
Green, Frank N., sheriff, N. Bostwick
Greensmith, Wm., mason, N. Wash.
Gelnet, Geo. W., ret far., S. Cochrane
Grier, Mrs. L. H., wid., W. Lawrence
Goble, Philip, wagons and carriages, W. Lawrence
Gundlock, Fred, laborer, S. Foote
Gunn, C. A., housewife, N. Cochrane
Gribble, Marion
Hart, Mrs. Jane, wid., W. Seminary
Hall, G. J., cuslder, W. Harris
Hall, Laura S., widow, W. Harris
Hall, Marcus, carpenter, Johnson
Haynes, William, mason, N. Cochrane
Hartell, J. C., butcher, N. Clinton
Hamilton, F. B., under sheriff, N. Cochrane
Hagerman & Barman, grocers, So. Cochrane
Hann, John, grocer, Forest
Hawley, Daniel, retired, N. Cochrane
Halpin, John, farmer, Foote
Harmon, Clyde E., clerk, N. Wash.
Hauge, Susannah, wid., N. Sheldon
Hart, Jeanette, h'sw'fe, N. Cochrane
Hart, H. J., carpenter, N. Cochrane
Hayden, John, tinner, W. Lawrence
Hayden, Celia, widow, W. Seminary
Harfort, C., wagon manuf'r, Pleasant
Haslett, J. C., clerk, So. Bostwick
Hatch, Vina
Hart, Catharine, widow, E. Henry
Harswell Luletta, widow, So. Sheldon
Hammond, Emma, hswfe, S. Sheldon
Haas, Lewis, billiards, So. Oliver
Hall, Ellick S., ret. mer., W. Seminary
Hayes, Tenryck, clerk, So. Bostwick
Henderson, Etta, E. Lovett
Hartwell, Seymore
Herbst & Son, tailors, S. Cochrane
Hayman, M., butcher, N. Sheldon
Holiway, H. W., bill'ds, So. Cochrane
Hill, Alonzo, carpenter, So. Sheldon
Hill, Benj, painter, So. Cochrane
Hill, Fred, painter, Pleasant
Hill, Elmer, drayman
Hines, Emeline L., hswfe, Washington
Hobbs & Co., T. D., gro., S. Cochrane
Holden, John G., farmer, Forest
Howe, M. H., carpenter, N. Cochrane
Hoss, Lucretia, hswfe, N. Washington
Houck, J. W., finisher, Stoddard
Hood, Eliza, widow,
Holbrook, L. C., w'r clerk, S. Sheldon
Henry, James
Hough, Charlotte, ho'sekpr E. Henry
Homer, Geo., Sr., laborer, Forest
Hauser, W. K., Forest
Hauser Geo. L., attorney, Forest
Henry, F. H., D. D. S., E. Lovett
Hotchkiss, S. A., barber, So. Cochrane
Huggett, Geo., attorney, E. Harris
Huffman, Glen
Holtz, Peter, butcher, So. Cochrane
Huber, Daniel, farmer, W. Lawrence
Ives, Andrew J., Vice President bank, E. Lovett
Jaskle, Jacob, gunsmith, S. Cochrane
Jackson, Andrew, farmer, S. Cochrane
Jackson John, organs and pianos, S. Cochrane
Jenny, Marietta, widow, N. Bostwick
Jennings, C. M., hardware, S. Cochrane
Jennings, H. K., Cashier Merchants National Bank, So. Cochrane
Jenkins, Dor, barber, So. Cochrane
Jessup, Geo., pensioner, Lansing
Jones, D. M., painter, E. Lovett
Jewett, Cyrus, carpenter
Jipson, William, laborer, Foote
Johnson, James, constable, E. Harris
Jones, E. C., supplies, So. Cochrane
Johnson, Mrs. T. L., ho'sewife, Clinton
Johnson, Mrs. Horace, So. Bostwick
Johnson, C. S., teamster, Forest
Johnson, Jo, laborer, N. Sheldon
Johnson, Jordan, farmer, S. Cochrane
Jones, William, mason, Shaw
Jones, Sylvia, widow, John
Jones, John, janitor, Oliver
Jordan, William, grocer
Johnson, P. H., ret farmer, E. Henry
Johnson, C. D.
Katz, Mathew, finisher, W. Prospect
Kauffman, Benj., laborer, Foote
Kauffman, Peter, taylor, Pleasant
Kellogg, Bradford, mason, W. Harris

CITY OF CHARLOTTE.

Kessler, A. H., lumber, W. Harris
Ketcham, Seth, supt., W. Lawrence
Ketcham, M. P., h'swife, W. Lawrence
Ketcham, Myrtle C., artist, W. Lawrence
Kennedy, Marie J., housewife, W. Seminary.
King, Lorenzo
Kinne, Malintha, milliner, S. Cochrane
Kiplinger, James, laborer, Johnson
Kilbourn, F. H., retired, N. Bostwick
Kimberly, Mary, housewife, Pearl
Kimberly, Erastus, retired, Pearl
Kirkland, E. P., ret. farmer, Prairie
Kimmel, Isiah, laborer, Lansing
Kies, A. H., laborer, Lansing
Kiplinger, Grant, laborer, High
Kiefenbauer, C., laborer, Foote
Kirkland, Geo., shoes, So. Cochrane
Kleine, William, laborer, N. Clinton
Kleine, Emanuel, retired, N. Sheldon
Klock, J. D., cigar m'n'r, So. Sheldon
Klock, Peter, painter, N. Washington
Knowles, Fred, engineer, Clinton
Krebs, J. S., sewing machines and organs, So. Cochrane
Lacey, W. P., Cashier First National Bank, W. Harris
Laugest, L. A., laborer, N. Washington
Laugest, Wm. laborer N. Washington
Lamb, G. B., grocer, W. Stoddard
Lamb, G. H., cabinetmaker, Lincoln
Lamb, E. J., housewife, Lincoln
Lamb & Sueheer, grocers, S. Cochrane
Lang & Vosberry, etc., S. Cochrane
Laughery, Isaac, laborer, W. Henry
Lane, L. N., S. D. A. officer, Pearl
Lambert, C. G., carpenter, E. Henry
Lampson, S. harness, So. Cochrane
Lounon, Martin, drayman, Railroad
Lewis, B. A. Mrs., widow, So. Clinton
Le Rose & Co., fruits, So. Cochrane
Lee, Geo. D., M. E. minister, Battle Creek Road
Levy, John, clothier, W. Seminary
Lightner, John, farmer, Lansing
Light, Geo. F., shoem'k'r, N. Cochrane
Lightner, Priscilla, h'swife, Lansing
Lisby, Ruth, widow, S. Sheldon
Lockard, Adaline, housewife, S. Cochrane
Long, Martha J., widow, E. Lawrence
Loring, A. T., real estate, N. Clinton
Loop, Albert, carpenter, John
Lasee, Levi, laborer, Pearl
Love, Lucy, widow, E. Henry
Loveland, Fred. H., dry goods, S. Cochrane
Lovell, Olive, h's'ekeeper, S. Sheldon
Lowmaster, Wm. carp'ter, S. Sheldon
Ludwig, L. J., laborer, Pleasant
Lynd, E. H., barber, W. Henry
Marple, F. D. restaurant, W. Harris
Mason, C., sea captain, N. Bostwick
Martin, R. T., ret. farmer, N. Clinton
Markham, O. J., ret. frm'r, S. Sheldon
Marks, John, painter, Lincoln
Matson, Wilbur, ret. farmer, McClure
Matson, Roswell, ret. far., Pleasant
May, John H., trav. salm., S. Cochrane
Mason, J. D., laborer, Prairie
Masten, C. M., farmer, N. Sheldon
Martin, Amanda M., housewife, W. Lawrence
Maynard, H. S., pros att'y, W. Harris
Mathews, Bert, foreman Bennett's factory, W. Henry
Merker, D. W., ret. far., So. Cochrane
Merrill, John W., laborer, N. Oliver
Merrill, Sarah, ho'sekeeper, N. Oliver
Merritt, Mrs. P. E., housekeeper, W. Harris
Merritt, C. A., supt. poor, W. Harris
Merritt, C. J., liveryman W. Harris
Mitchell, Herod, lab'r, W. Seminary
Miller, Levi, laborer, So. Cochrane
McConnell, James, retired, Lansing
Merritt & McClintic, dgs, S. Cochrane
Merritt, Frank, M. D., Pleasant
Merritt, William A., foreman in Curtis factory, S. Sheldon
Metzinger, William, billiards, N. Oliver
Metzinger, E. B.
Merchant, D. J., laborer, So. Sheldon
Meyers, J. G. carpenter, So. Oliver
Miller, Jane M., housewife, W. Henry
Miller, John G., agriculture, W. Henry
Miller, John L., harnessm'k'r, Prairie
Miller, M. K., foundryman, N. Cochrane
Miller, John A., foundryman N. Cochrane
Miller, Andrew J., laborer, N. Cochrane
Miller, John H., laborer, S. Cochrane
Mikesell, Jessie, cannery, W. Seminary
Mikesell, L. P., cannery, S. Sheldon
Mikesell, M. J., dressm'k'r, S. Bostwick
Mikesell, Ablinda, widow, W. Henry
Mikesell, Mrs. William P., widow S. Sheldon
Mists, Mrs. Frances, h'swfe, N. Sheldon
Michel, Jacob, cigar m'f'g'r, Cochrane
Mitchell, D. M., butcher, So. Main

Mikesell, Jacob, farmer, W. Prospect
Monk, William, retired, N. Cochrane
Moll, Henry, cooper, Railroad
Moll, Geo., laborer, Robinson
Mosier, Jackson, clerk, W. Lovett
Moore, A. B., patent rights, N. Wash.
Morey, Addie C., teacher, N. Oliver
Morse, H. F., widow, W. Henry
Moomey, James A., farmer E. Henry
Mount, Sylvia, widow, E. Henry
Morgan, C. J., laborer, Forrest
Morrison, Henry, retired, W. Henry
Moyer, H. A., ret. farmer, S. Cochrane
Morey, D. E., clerk, N. Cochrane
Mosell, Otto, cabinetm'k'r, Stoddard
Mulholland, W., ret. farmer, Beech
Munger, J. W., hardware, E. Henry
Munson, Mrs. A. H., wid., S. Cochrane
Munson, Almira, widow, N. Bostwick
Murray, Albert, boots and shoes, E. Lawrence
Mygrants, Jacob, retired, N. Cochrane
Myers, Abigail, widow, E. Henry
McArthur, Mrs., widow, Shaw
McCartney, J. M., farmer, W. Harris
McCaskey, Geo., laborer, Forrest
McCargar, William, far., N. Cochrane
McCall, Jennie, h'swife, N. Cochrane
McCall, L. H., attorney, N. Cochrane
McClennon, Melvina, wid., W. Lawrence
McClellan, Maggie, h'swife, E. Lovett
McConnaughy, William, mason, So. Cochrane
McCarroll Bros., steam laundry
McClure, C. H., grist mill, N. Cochrane
McClintock, Lovina, widow, E. Henry
McCoy, Orange, ret. milliner, S. Oliver
McDougall, W. M., c'nst'r, N. Cochrane
McDonald, Grace, music teacher, E. Lovett
McDonald, Frank, st'kbuy'r, E. Henry
McGrath, Fred, grocer, So. Sheldon
McGrath, H. T., printer, So. Sheldon
McIntyre, James, ret. far., S. Cochrane
McGrath, Margareth, wid., S. Cochrane
McLaughlin, Rebecca, artist, Prairie
McPeek, J. L., Judge of Probate, N. Cochrane
McUmber, C. H., farmer, W. Shaw
Newton, Susan, h'swife, S. Cochrane
Newton, blacksmith, N. Sheldon
Newton, Mrs. L. L., wid., W. Seminary
Newland, M. D., So. Cochrane
Newcomb, M. E., drain com., W. Shaw
Nichols, J. J., laborer, Forest
Nichols, J. C., attorney, N. Cochrane
Nichols, Stephen, lab'r, W. Seminary
Nichols, A. L., coll'ct'g ag'nt, E. Lovett
Nichols, Edgar F., laborer, High
Ober, Chas., laborer, Brackett
Odell, Aaron, W. Henry
O'Bryon, Mrs. Wm., widow, Prairie
O'Neil, Edward, saloonist, N. Oliver
Owings, F. H., farmer, W. Prospect
Owings, Georgia A., wid., Johnson
Olmstead, Otis, laborer, N. Sheldon
Olmstead, O., laborer, N. Clinton
O'Leary, C. R., milk, W. Lawrence
Packard, O. E., grain, E. Lawrence
Packard, Eliza, h'swfe, E. Lawrence
Patterson, P. D., M. D., E. Lawrence
Patterson, E. J., jeweler, So. Clinton
Parrish, Alex, laborer, E. Foote
Parrish, Homer, laborer, E. Foote
Paine, Henry, laborer, Forest
Palmer, Stephen
Palmer, F. C., M. D., So. Cochrane
Palmeter, Mary
Parkhurst, J. D., retired, N. Sheldon
Parker, Sarah A., wid., N. Cochrane
Perkins, Mariah, housewife, John
Perkins, John
Perry, Geo. A. Publisher, W. Henry
Peskett, Geo., sec'for'n, N. Washington
Perry, Chas. A., well digger, Shaw
Perry, Mrs. Joseph, wid., S. Cochrane
Phillips, M. S., D. D. S., N. Sheldon
Pierce, T. C., farmer, Stoddard
Pixley, Lucy M., prop. Phelps house, N. Cochrane
Pierson, Helen L., h'swfe, S. Cochrane
Pike, F. W., showman, E. Henry
Piper, Wm., retired, W. Lawrence
Pitcher, Chas., teamster, E. Shaw
Pitcher, Elizabeth, wid., So. Sheldon
Parish, James, laborer, Lansing
Parish, Perry
Pangborn, Chas., carpenter, Lansing
Paine, Fred, laborer, Forest
Pangler, C. H., farmer, Beech
Proctor, Herbert, laborer, High
Preston, E. A., gardener, Pearl
Pitcher, Edwin, mason, E. Shaw
Platt, Elbert, laborer, N. Sheldon
Pollock, Samuel, retired, Park place
Pollard, James G., bookkeeper, N. Cochrane
Polhamus, Emma, h'swfe, Lincoln
Potter, John C., grocer and manufacturer, Stoddard
Potter, F. H., oper'r G. T., N. Sheldon
Potter, Sarah L., widow, Pearl
Powers, Hannah J., h'swfe, Lansing

Porter, Geo. A., carpenter, E. Foote
Powers, J. D., retired, Lincoln
Powers, Lavina, wid., S. Washington
Powers, J. M., attorney, Warren
Powell, J. W., farmer, N. Forest
Pratt, Mrs. J. M., wid., W. Lawrence
Preston, Adaline M., h'sfwe, Prairie
Priddle, Malinda, widow, E. Henry
Prince, C. E., clk Wilson & Spaulding
Rallie, Alice, h'swfe, So. Cochrane
Rand, Agnes, widow, N. Cochrane
Rand, W. H., M. D., N. Cochrane
Ramy, N. C., clerk, So. Washington
Ramy, Alex, carpenter, Lansing
Ramy, J. H., farmer, Lansing
Rathburn, Lora, housewife, High
Rathburn, S. B., laborer, High
Randall, Benj., foreman Charlotte Mfg Co. shops
Randall, C. H., laborer, N. Sheldon
Ramer, Dora, h'swfe, N. Bostwick
Resch, Wm., carpenter, Stoddard
Reynolds Bros., dry g'ds, S. Cochrane
Reed, Geo. W., carpenter, Lansing
Robfus, Gottfred, blksmith, Lansing
Richardson, G. W., engineer water works, N. Cochrane
Rice, Althea, widow, N. Washington
Riley, D. J., laborer, N. Cochrane
Richardson, D. J., laborer, S. Sheldon
Richardson, J. J. ret. clo., S. Cochrane
Rickard, Mrs. Daniel, wid., S. Sheldon
Rickard, John E., teamster, John
Robinson, Chas., baggagemaster G. T. R. R., McClure
Robinson, Samuel, ret., E. Lawrence
Rowley, Geo. W., insurance and real estate, E. Lawrence
Rowley, Lafayette, bookkeeper, E. Lawrence
Roberts, J. C., blacksmith, E. Shaw
Robb, Mrs. M. W., widow, W. Harris
Robbins, C. V., car. trim., So. Cochrane
Rowe, Julia N., h'swfe, W. Henry
Rogers, Henriette, wid., S. Cochrane
Rogers, Clara, h'swfe, So. Cochrane
Rose, Nathan H., carp'ntr, N. Sheldon
Rullison, T. C., blacksmith, Lansing
Rickard, Wm. W., laborer, S. Sheldon
Sanderson, J. B., retired, Foote
Sentor, J. L., ret. farmer, So. Sheldon
Sawyer, Moses, ret. far., So. Sheldon
Scott, Anson, ret. farmer, N. Oliver
Selkirk, A. N., jeweler, E. Lawrence
Searles, Phebe E., wid., E. Seminary
Senson, Sarah D., bskpr, N. Oliver
Seavolt, W. S., travelling salesman, Pleasant
Sattler, Geo., ret. farmer, Pleasant
Shepherd, Elisha, Sr., far. S. Cochrane
Shepherd, James, drggist, N. Cochrane
Shepherd, L. H., grain, N. Cochrane
Shepherd, Emeline, bskpr, S. Cochrane
Shepherd, Mrs. A. L., wid., W. Seminary
Sherwood, Geo. W., J. P., S. Cochrane
Shaull, Warren, ret., So. Cochrane
Shaw, Joseph, far., N. Washington
Shear, John, farmer, Lansing
Shriner, Mrs. M. W., artist, S. Cochrane
Shroads, James, laborer, N. Clinton
Shuler, O. P., druggist, So. Bostwick
Siesler, A. H., wall paper, E. Lovett
Smith, J. M. C. attorney, Warren av.
Smith, Chas., retired, Clinton
Smith, Harriett, h'swfe, So. Sheldon
Smith, N. L., manager Charlotte Manufacturing Co., So. Sheldon
Smith, Richard, mason, W. Lovett
Smith, Louis J., clerk, W. Lovett
Smith, Adelbert, laborer, N. Cochrane
Smith, Mrs. Chas., wid., W. Seminary
Smith, L. G., abstractor, N. Bostwick
Smith, Mary J., h'swfe, N. Bostwick
Smith, F. G., city treas., W. Harris
Smith, J. B. & Co., hsu. Horatio
Snow, T. D., laborer, So. Sheldon
Snow, Newton, farmer, So. Cochrane
Snyder, Geo. A., farmer, Lincoln
Snyder, Chas., laborer, W. Lawrence
Snyder, J. F., carpenter, Shaw
Snell, L. A., M. D., So. Cochrane
Snyder, Diana
Snyder, N. S., laborer, So. Sheldon
Snyder, Geo. A., farmer, Lincoln
Snyder, Robert
Solomon, Mary A., h'swfe, N. Oliver
Solomon, A. A., mason, N. Washington
Solomon, Isaac, laborer, Amity
Solomon, Samuel, laborer, Amity
Sprouse, E., laborer, Lansing
Spencer, Geo. H., grocer, S. Clinton
Spencer, Mary J., wid., W. Lawrence
Spreng, Ada, Prop. Peninsular House,
Sprague, Mrs. Z. H., wid., S. Cochrane
Sprague, Gideon
Spaulding, Mary M., widow, W. Henry
Spaulding, A. P., boots and shoes, E. Henry
Spaulding, Frank, agricultural implements, S. Cochrane
Sprouse, S. B.
Squier, J. H., retired, W. Lawrence
Squier, Geo. W., retired, N. Sheldon

Stealy, Eva, housewife, Pleasant
Stealy, A. R., M. D., Pleasant
Stoner, Henry, mason, John
Stoner, J. C., merchants, So. Sheldon
Stamp, Lafayette, clerk, Warren
Stone, housewife, W. Lovett
Stone, Calvin, janitor Congregational church, W. Lovett
Stone, L. A., harness store, S. Cochrane
Stockwell, Cyrus, preacher, W. Seminary
Strother, J. T., boots and shoes, W. Lawrence
Stewart, Nelson, carpenter, S. Clinton
Stevens, Ann C., widow, S. Cochrane
Stewart, Benjamin, retired, Lansing
Stewart, R., news dealer, W. Henry
Sisson, D. F., mason, Amity
Sweet, A. J., mayor, W. Henry
Sweeney, John, laborer, W. Henry
Stocking, C. H., carpntr, W. Lawrence
Stall, Jennie S., housewife, S. Wash.
Sommers, Geo., engineer, S. Cochrane
Tanner, William, ret. far., N. Clinton
Taggart, C. H., engineer, W. Lawrence
Taylor, Chas. E., marshal, S. Sheldon
Thomas, J. Q., retired, W. Lawrence
Thomas, P. M., trav. salesman, E. Seminary
Thomas, Horace S., farmer, Shaw
Thomas, H. M., farmer, S. Cochrane
Thompson, Chas. O., street sprinkler, N. Sheldon
Thompson, Frank, laborer, N. Clinton
Thompson, Sarah E., housewife, N. Clinton
Thorpe, Wilson, carpenter, S. Sheldon
Tirrill, J. F., ret. farmer, E. Harris
Titus, W. S. Jr., trav. salesman, N. Cochrane
Titus, W. S., C. B. minister, S. Oliver
Tracey, Abigial, h'swife, N. Cochrane
Tracey, William, teamster, Pearl
Tremaine, D. L., livery and hack line, E. Lawrence
Tobias, Louisa, widow, W. Stoddard
Turner, L. H., grocer, E. Seminary
Tucker, Jennie, dressmaker, W. Henry
Twogood, S. B., carpenter, N. Sheldon
Twitchell, S. N., ret. far., E. Henry
Underwood & Lyud, barbers, So. Cochrane
Van Horn, C., laborer, Forest
Vanderhoof, Peter, ret. far., W. Lovett
Vanderhoof, lumber ag't, S. Sheldon
Vantassel, V. T., farmer, W. Prospect
Vantassel, O. L., farmer, W. Prospect
Vickery, Reuben, night watch, N. Clinton
Vickery, Daniel, plumber, E. Henry
Walrath, W. S., carpenter, W. Foote
Wakendorf, C. R., plumber, E. Lovett
Wells, Mary J., widow, Merritt
Weston, Yet, laborer, S. Cochrane
Whitton, H. C., retired, E. Lovett
Whitton, Fred
Wood, C. H., farmer, Beech
Waltersdorf, Frances E., wid., N. Sheldon
Waller, W. H., blacksmith, N. Clinton
Wareham, Philip, bag. agt. M. C. R. R., N. Sheldon
Warner, Margaret, h'swife, Battle Creek
Warner, Otis, laborer, Battle Creek
Walker, R. L.
Warren, R. W., sinking, deeds, S. Bostwick
Warren, M. B., liveryman, S. Sheldon
Wholihan, John T., day operator M. C. R. R., Stoddard
Warren, Henry R., ret. far., W. Henry
Walker, Effie, housewife, S. Sheldon
Walker, W. H., ret. far., W. Seminary
Walters, A. C., agricultr'l imp., W. Henry
Ward, Frank W., grocer, W. Henry
Ward, Nancy N., milliner, E. Lawrence
Warren, F. G., milliner on P. M., S. Cochrane
Welch, Hiram, drayman, N. Clinton
Welch, Frances, widow, W. Henry
Weeks, M. B., laborer, N. Sheldon
Wehler, D. F., J. P., W. Henry
Weaver, F. A., M. D., E. Seminary
Webster, Cobb & Co., lumber, W. Lovett
Wetmore, Eunice, widow, S. Oliver
Wilcox, Horace, far., ex Reg. deeds, N. Clinton
Wilton, L. W., farmer and produce, S. Cochrane
Winslow, H. H., widow, S. Sheldon
Williams, Charlotte, wid., W. Seminary
Winchell, N. D., laborer, W. Henry
Wilders, Job, crockery and glassware, S. Cochrane
Williams, Geo. A., livery, E. Henry.
Wildt, Chas., ret. farmer, E. Henry
Whaley, Ezra, retired, W. Harris
Wheaton, R. H., bridge builder E. Lovett
Wheaton, Chas., bridge builder, Prairie
Whitford, Hattie, h'swife, E. Lawrence
Whitaker, Flora J.,
White, Addie, widow, S. Cochrane
Whitright, R. J., clerk, S. Sheldon
Wilson, J. T., Agr. Imp., E. Lawrence
Wilson & Spaulding, Agr. Imp., Lawr.

Wilcox, Jerusha P., widow, N. Cochrane
Williams, W. B., Cong. minister, W. Stoddard
Winters, Josephine, widow, Summit
Whitcomb, Lucy, housewife, Summit
Worden, Mrs. M. K., widow, N. Oliver
Wood, Mrs. N. G., widow, S. Oliver
Woodberry, Almira, housewife, Prairie
Woodberry, Geo., plumber, Prairie
Wolston, Joshua, laborer, E. Shaw
Young, E. D., manufacturer, Forest
Zimmerman, S., laborer, Forest

GRAND LEDGE—FIRST WARD.

Aldrich, Henrietta
Archer, Carlton
Aldrich, Chas.
Alexander, Camina
Aldrich, Frank
Allen, Clarence
Arms, F. C.
Astley, Ed.
Astley, Jo
Allen, E. P.
Bowen, James
Bardick, Mrs. C. D.
Barber, W. A.
Bosworth, Reason
Brenger, Joseph
Bailey, Sol
Bongaman, Elmer
Barber, John
Billings, W. B.
Beach, Fletcher
Babcock, Dora
Bellis, Mrs. Levi
Blanchard, M.,
Bullard, D. C.
Berry, Geo. N.
Bailey, Myron
Bigelow, Emma
Blaine, Peter
Beach, M. P.
Bailey, Geo. H.
Babcock, W., J.
Bongamyn, Orrin
Blanchard, Hannah
Blanchard Frank
Brown, Geo. A.
Bartch J. M.
Brown, Nettie E.
Brinkerhoof, Lola
Beck, C.
Brown, D. D. F.
Beach, Emily
Brown, Ambrose
Bolder, Tom
Briggs, E. M.
Chilson, Jas.
Clark W. H.
Cole, Mrs. Ezra
Capwell, Wm.
Cole, Harriett
Chadwick, Nancy
Campwell, Geo. W.
Campwell, J. M.
Carver, Watson
Campwell, Milo
Capwell, John
Coryell, Geo. L.
Cole, D. E.
Cramer, Eugene
Chamberlin, Jas.
Campbell, A. E.
Cole, Emily
Cronk, John
Dennis, Samuel
Deitraff, Mrs. E. M.
Dravenstatt, Harrison
Derbyshire, Mrs. J. E.
DeCou, Mrs. Bettie
DeCou, Mrs. Anna
Deitraff, Emanuel Est.
Deryea, Samuel
DeWitt Al
Dutcher, David
Dudley, M. P.
DeWitt, J. W.
DeWitt, Jas. N.
Earl Emily
Knox, B. M.
Edwards, Bedman
Earl Johanna
Earl, Helt
Earl, B. F.
Earl, Abiey
Edwards, Letty
Earl, C. C.
Ealer, Minerva
Fritz, D. C.
Foreman, Tobias
Fuller, Fred
Frickert, Geo.
Fitzgerald, J. W.
Frees, Frank
Foster, Sanford
Foreman, D. C.
Granger, Mrs. S. B.
Grenter, Orin
Grand Ledge Sewer Pipe Co.
Grand Ledge Canning Co.
Green, G. B.,
Gillam, Maldon

GRAND LEDGE.—1ST WARD.

Grand Ledge Building Association
Glenn, Chris
Grand Ledge Chair Company
Gunsaulbauser, M. B.
Griswold, Emerson
Goodale, H. K.
Gates, Samuel
Hall, Wm. H.
Hixson, A. D.
Hixson, E. J.
Holmes, J. S. H.
Holmes, J. S.
Halliday, Oliver
Henderson, John
Hayward, Estella
Hart, Ben
Hilliker, T. W.
Herrington, Eli
Howe, Judd
Hudson, V. E.
Hemlock Point Association
Heart, James
Hosher, Howard
Hulce, Hannah
Hurley, Lucy V.
Huddleston, Isabell
Hones, Harriett C.
Haven, Mrs. Wm.
Haggart, A. J.
Hulce, Ben
Hewett, S. P.
Holmes, J. C.
Holmes, N. W.
Holmes, Mary L.
Hixson, Herbert
Hoag, V. W.
Holmes, L. W.
Holmes, Hattie
Hixson, Abram
Hixson, Arnson
Hixson, Mrs. Rose
Holmes, Amos
Hixson, Fred
Hilliker, Charles
Holmes, Harley
Homer, A. L.
Jenne, Silas
Jenne, Christina
Johnson, Hiram
Johnston, Milford
Jones, Mrs. Chloe
Kerr & Clark
Kiser, Al
Klumpp, Herman
Kennedy, G. D.
Kennedy, Bert
Kent, D. L.
Kent, V. M.
Kent, C. A.
Kent, E. G.
Kiser, L. J.
Kehler, Chris
Krapp, Mrs. K. M.
Krapp, Frank
Kleinsmith, A. W.
King, P. A.
Lamkin, Clark & Hulce
Lamkin, Al
Laughton, J. C.
Laudenberger
Love, Ralph
Long, James
Loveless, Geo. B.
Loveless, Phebe A.
Lee, Amy
Lund, F. O.
Mascho, Geo. A.
Messenger, Mrs. S. S.
McGibbons, Wm.
Mannwarring, Wm.
Maier, Martin
McDowell, C. G.
Murphy, Albert
Miller, S. J.
McGowan, Samuel
Mascho, Almond
Mascho Bros.
McGibbons, Hattie
Middleton, Maria
Middleton, Ed
McPeek, Lydia
Morgan, John
Mudge, S. S.
Marshall, Joseph
Murphy, S. R.
McRoberts, Wm.
McCormack, N. J.
McBride, Mary A.
Newton, R. L.
Nostrant, Geo. W.
Nostrant, Mrs. Wade
Nixon, Mrs. Francis
Newman, Fred
Newman, Chas.
Niles, John
Otto, Joseph
Olmsted, Mrs. E. J.
Pearce, O. H.
Pratt & Son
Pownall, Volney
Persons, Frank
Phillips, C. H.
Potter, S. V., Jr.
Potter, H. C. S.
Peterson, Mrs. Etta
Parker, David
Parker, Rufus
Patterson, Earl
Porter, Henry
Russell, A. E.
Reynolds, Oscar
Bowman, Hugh
Rohl, John
Reed, James
Reed, Frank L.
Reed, Martha
Russell, J. F.
Russell, Calvin
Russell, Frank
Russell, Edmund
Russell, Selah
Rawson, F.
Reynolds, Clarissa
Reynolds, Alta
Reed, Chas. P.
Robinson, T. D.
Robinson, Sam'l
Russell, Wm. P.
Rogers, Mary M.
Rall, Lucy
Rogers, Thos.
Ramsberger, J. C.
Reed, Mert
Russell, J. N.
Rosier, Wolcott
Shafer, Jas.
Skinner, Wm. N.
Sebert, Laura
Spears, Simeon
Spencer, Jas.
Spencer, Milford
Shane, D. D.
Shaw, Eunice
Sackett, Chas.
Snyder, L. J.
Smith, David
Smith, J. W.
Searls, Lewis
Streeter, Anna J.
Streeter, J. W.
Streeter & Son
Streeter, M. T.
Streeter, E. J.
Stanton & Son
Shaw, J. W.
Summers, J. D.
Sickles, J. W.
Sprague, A. B.
Sebring & Co.
Shaddock, D.
Smith, Maria
Swagler, Henry
Simpson, Lester
Simpson, John
Sanders, David
Saxton, John
Schumacher, A. B.
Steele, W. B.
Salyer, John
Sitzer, F. A.
Sebastians, Mrs. S.
Sutherland, Hattie
Seidler, G.
Sheets, Geo.
Sanders, Elijah
Smith, Robert
Smith, Mrs. G. B.
Spencer, Ernest
Smith, Byron
Smith, Angeline
Tester, Mrs. Adell
Traver, M. S.
Taber, A. J.
Titcham, Wm.
Tater, Frank A.
Taylor, Mrs. Ed
Tous, Wm.
Timmerlin, Mrs. Emma
Tibbetts, Elizabeth
Thomas, Geo. W.
Thomas, A. J.
Taylor, J. P.
Townsend, F. G.
Townsend, Mrs. S. A.
Tucker, J. C.
Vanalbo, Mrs. N. M.
Van Iterzilgen, Peter
Valentine, Ben
Wareham, H.
Wareham, J. C.
Walds, L. P.
Westland, W. C.
Watson, N. G.
White, Sam
Warner, J. L.
Wilkinson, B. F.
Wilkinson, M. L.
Wright, Wm.
Wright, Fred
Wright, Elbridge
Whelpley, S. H.
Williams, Mark
Wickwire, Ellen
Wickwire, Nelson
Wythe, W. T.
Wythe, Mary
Wythe, Chas.
Wickler, W.
Wallace, Mary
Wells, Ralph
West, Rose
Wilson, W. E.
Zufelt, Juliette

2ND WARD.

Albright, A.
Ames, Willis
Astley, E. C. & Son
Abbey, H. L.
Appleton, W. T.
Austin, James
Arbor Hall Association
Briggs, George
Boyle, Fargo
Baker, Geo.
Blomberg, Mrs. M.
Brown, Geo. W.
Brown, W. H.
Benjamin, Chas.
Beach, M. P.
Bartch, J. M.
Bartch, Martha
Bigelow, Henry
Brown, G. J.
Dowell, John
Wyche, John F.
Brown, S. A.
Bedford, H. D.
Bennett, F. P.
Bigelow, J. H.
Barnum, A. P.
Bancroft, John
Hale, Henry
Bush, C. H.
Covey, Sarah A.
Covey, C. J.
Clayton, P. P.
Cannon, Chas.
Campbell, S. P.
Cummings, A. S.
Case, J. B.
Clark, Gilbert
Chapman, L. W.
Chesmond, Wm.
Chapman, Clark
Chapman, L. D.
DeWitt, Jacob N.
DeWitt, Mrs. Jonathan
Davis, W. R.
Davis, A. C.
Davis, M. O.
DeWitt, Geo. H.
DePuy, Fayette
Detweiler, Isaac
Dilmaster, Geo.
Dunn, Thos.
Daniels, Wm.
Dean, Cyrus
Doty, Nelson
Ehler, Mrs. Edward
Eaton, J. P.
Eley, Robert
Eaton, Robert
Foster, Albert
Foster, Frank
Flemming, C. W.
Fuller, S. L.
Field, Wm. H.
Follett, Eugene
Fisher, Martin
Foster, Eugene
Fritz, J. F.
Goodale, John
Gillam, Mahlon
Girch, Amy
Gilmore, Lewis J.
Griswold, Emerson
Glover, Mrs. Louisa
Green, Walter
Gates, Leonard L.
Gilmore, Alfred
Gilmore, Wm.
Hollis, Mary
Howard, Robert
Hyde, W. G.
Henderson, Robert
Halstead, J. E.
Hixson, E. J.
Hall, Mrs. Charles
Howe, Joshua
Halbert, Mrs. Harrison
Himes, Charles
Holt, C. W.
Howard, Steven
Hummell, M. V.
Hodge, George
Irish, Geo. W.
Irish, Geo. E.
Johnson, Mrs. C. B.
Jerome, Susan
Jenkins, Thomas
Kimball, Jason
Ketchum, Sarah
Kebler, Chris
Keep, Clyde
Kenny, M. J.
Kenny, B. W.
Kenny, Elgar
Kelley, John
Keenan, Charles
Lanier, Lewis
Lewis, Hiram
Lovett, Amanda
Little, Geo. W.
Mattson, Henry
McKenney, Mrs. Wm.
McMillen, Wm.
Morley, S. L.
Meyers, Fred
Merrill, Mrs. Ed.
McKinn, Daniel
Metcalf, Wilbur
McBride, Mrs. Mary
Moon, Bert
Mudge, J. S.
Moulder, A.
Miller, Wm. H.
McPeek, Rose
Mitchell, Emma A.
Miller, C. L.
Moss, J. K.
Noyes Bros.
Noyes, Frank
Noyes, Ed.
Persons, Louis
Phillips, Perella
Phelps, Mary J.
Parker, Mrs. Fannie G.
Parker, Adam
Phillips, Grant
Pruden, J. B.
Pearce, M. L.
Parker, Betsy A.
Pearl, L.
Pike, James
Persons, Wm.
Quale, R. M.
Reed, A. H.
Rossman, Wm.
Rogers, L. P.
Ring, W. H.
Remley, F. F.
Ring, I. G.
Rosengole, E. E.
Rogers, Mary D.
Rich, Earl
Riley, Hiram
Rowland, Henry
Rowland, Daniel
Rogers, Mary
Roberts, Hutter & Co.
Seaburn, Frank
Sharp, Geo.
Strickland, S. A.
Swiger, Geo.
Smith, S. B.
Smith, H. W.
Swift, Frank
Shaddock, A. E.
Snyder, Emery
Summers & Wareham
Sloan, Nellie
Sheets, S. J.
Smith, George
Smith, David
Sutherland, M.
Story, Chas.
Smith, H. E.
Shaddock, Levi
Sutherland, Edson
Sly, Lester
Tong, Harriet
Toms, Thomas
Thomas, Tubbs & Co.
Talleman, A.
Tickham, W. H.
Tater & Spencer
Vanderlanger, Mrs. John
Van Wie, A.
Vandercolgen, J. C.
Vanderwelt, W. T.
Wright, Oliver
Williams, W. J.
Williams, Mrs. Geo. W.
Woolcuff, Mrs. C.
Williams, Wm.
Wellman, J. T.
Wood, C. M.
Wood, Reuben
Walser, Albertus
Winnie, James
West, Milford
West, M. L.
Ward, Fay
Walfrain, Mrs. Rosella
Watson, A.
Wheeler, Geo.
Wheelock, O.
Wheelock, Carl
Walker, G. W.
Walker, J. B.
Williams, Hannah
Whitneck, Rachel
Walters, John
Whitney, S. G.
Whitney, Ed
Wells, Earl
Young, Geo. A.
Young, Geo. W.

EATON RAPIDS CITY.

Allen, Mrs. H. M.
Adams, W. A.
Adams, Mrs. S. L.
Ashley, David
Anderson, Fayette
Andrews, Nelson
Anderson, G. N.
April, John
Annis, A. W.
Andursky, Sam
Abele, Frank
Allen, George L.
Andrews, John J.
Arnold, Ursula
Ayers, W. G.
Anderson, Mary L.
Baker, E. H.
Butterfield, L. L.
Bacon, Nettie
Braynard, F. M.
Bacon, Anna
Bemis, Catherine
Barnard, Mrs. M. J.
Brady, Sophia, Estate of
Brewer, Mrs. A. M.
Bandy, L.
Bursh, Thomas J.
Briggs, W. C.
Blake, Hester
Bissell, T. M. Plow Co.
Blair, G. B.
Botts, John
Bentley, S. A.
Bradt, Richard
Blacker, John
Barton, L. M.
Brooks, O.
Boatman, Jacob
Birney, Francis
Bateman, William Seward
Blodgett, Richard
Bradley, J. M.
Braynard, W. D.
Belmbustead, James J.
Bentley, L. A.
Benge, Mrs. F. B.
Bartley, George
Bellman, Catherine
Brown, Ida W.
Bruer, E. J.
Ball, A. P.
Bresselling, Mrs. T. J.
Blackmon, Mrs. F.
Britton, Henrietta
Bentley, C. B.
Bartlet, A. J.
Birney, Peter C.
Bostedor, J. H.
Boody, Mary
Buchanan, Lert
Blaire, G. B.
Beatty, William
Booth, D. L.
Bidwell, A. J.
Benge, Helen
Belknap, C. G.
Bostedor, James
Betzer, Henry
Belknap, Morris
Crawford & Sterling
Claflin, John D.
Crawford, M. D.
Cowen, G. O.
Cobb, William
Caldwell, E. D.
Crumley, Robert
Case, M. L.
Clark, M. L.
Caster, Chas.
Cowan, A. H.
Clark, John E., Est.
Clark & Griffin
Cronen, M.
Carper, Wm.
Corbin, E. D.
Cowles, Martha, Est.
Chapman, L. E.
Corbin & Harris
Cowden, James
Capron, H.
Carr, Joseph
Champlin, Mrs. Lyman T.
Cook, D. H.
Casler, Wesley
Crusoe, Benj.
Clark, P., Est.
Clark, William
Clay, Frank M.
Cook, Mrs. J. A.
Covey, Mrs. Benj.
Cox, Walter
Cole, Mattie J.
Connell, Marcus
Corbin, J. M.
Darling, Andrew
Demming, F. P.
Drewery, D.
Decker, Elizabeth
Devering, H. E.
Derby, Mrs. H. B.
Dearin, H. E.

1

EATON RAPIDS CITY.

Decker, Oscar
DeCourcey, Jess
Decker, Mrs B.
Decker, L.
DeGolia, F. H.
DeCourcey, Jerome
Dally, C. W.
Daniels, Mrs T. W.
Dutton, Allen C.
Dishro, Phineas
Davis, N. H.
Durand, Chas
Dialso, Lacey
Diabro, Bert L.
Dunbar, Jennie
Davis, Simon
Decker, Mrs. Jacob
Darling, Albert
Drake, Olive
Dewber, Irvin
Drake, Mrs. H. N
Davidson, B. E.
Dunlap, Annie E.
Elliott, Marcellus
Ediot, Geo. W.
Elder, Elias
Fay, J. H.
Field, L. H.
Ford, W. A.
Ford, Geo. W.
Fancer, Seth
Fuller, Geo. A.
Fassett, W. D.
Frost, Mrs. E. B., Est.
Fix, Mary
Fist, Caroline
Flynn, J. J.
Freer, Amanda
Fairchild, James
Fuller, S. B.
Frost, Frank
Fowler, L.
Fish, Jane
French, Mrs. Sarah
Filley, Caroline
Foote, A. G.
Field, Rev L. H.
Ferris, James
Ferris, Sarah
Fairfield, C. T.
Gary Bros.
Gillett, Fannie A.
Goodrich, A. D.
Griffith, Geo. P.
Green, Albert
Green, Clark
Grubb, Juliette
Geer, Rachell
Gould, Mrs A. B.
Giddings, Sarah
Goodrich, S. F.
Gaylord, Mary E.
Griffin, M. M.
Geer, Nellie
Green, Nettie
Gale, John G.
Godfrey, Mrs. Dolly
Gould, Margaret
Greer, Ella
Greer, Purcell
Gaylord, Mary C.
Garfield, Mary
Gould, Dan W.
Goodyear, Andrew
Gaylord, Mary E.
Garry, E. S.
Gallery Jack, Est.
Gallery, J. H.
Gifford, Chas E.
Gallery, Elma M.
Gallery, J. M.
Gallery, Emma
Gallery, Arthur
Hicks, Leonard
Harmon, Eugene
Horner, Samuel
Hess, Mrs. Hannah
Halbert, Azariah
Hicks, Liberty
Hudson, Mrs S. F.
Holmes, W. F.
Huntington, L. B.
Hamlin, Mrs. Ada L.
Hyde, Rufus J.
Hall, Belle
Hedley, Orval
Hartson, Ursula
Hill, Mary A.
Harris, E. S.
Hall, Edward S.
Holt, Wm. A.
Hyde, J. R.
Hall, Jesse
Hunt, Fanny
Harrington, S. R.
Havens, Sarah A
Holmes, J. J.
Hurd, Alton
Harris, Geo. E.
Hall, Betsy
Harmon, Becky
Henderson, John
Harwood, A. Est.
Henderson, W. S.
Hobart, Mrs. Anna
Hard, Marcus
Harris, Chas.
Hadley, Richard
Hall, J. T.
Hicks, George W.
Hunger, Mary A.
Hicks, Mrs. Lovina
Horner Bros.
Harper, Edwin
Hamlin, Mrs. John
Hamilton, H. H.
Hamlin, Frank
Huntington, Wm. J.
Humeston, James
Haner, Nelson
Humeston, Amanda
Haight, Seward
Hestler, D. R.
Humeston, Julius
Holmes, Eunice
Hall, Mary A.
Heslie, Miss N.
Harwood, Susan
Horton, Susan
Hames, J. H. Est.
Hamlin, Ada L.
Higgins, Wm. R.
Higgins, Eliza
Hicks, J. M.
Hammond, Daniel
Haight, Jacob
Haywood, B.
Huffman, Marian
Holmes, E.
Jopp, Nancy
Jacobs, T. H.
Jenne, Mrs. M. A.
Jordan, Mrs. William
Jackson, C. H.
Johnson & Wheeler
Johnson, Hiram
Jones, John
Kiper, M.
Kong, William
Keich, L. L.
Knight, Mrs Geo. W.
Knight, Amos
Knoblock, Mat
Koelsch, H. Bros.
Kissler, Adam
Knapp, E. F.
Knapp, George W.
Knapp, Sarah
LeFever, F. B.
Lindsley, David
Lindsey, Frank
Lewis, B.
Lewis, M. B.
LeFever, George
Leonard, Celestia
Latham, William
Lawrence, Ira N.
Lewis, C. N.
Long, S. A.
Lettis, C. M.
Lewis, O. B.
Lown, Hosea
Lincoln, Sarah N.
Leonard, Frank
Leisenring, Mrs. S. E.
Leisenring, John
Leisenring, P. C.
Lewis, Mary
Lane, O. W.
Leggett, Wm.
Leonard, Philip
Manning, Jerry
Markham, O.
Mester, Mrs. S. P.
Mester, Mrs. M. C.
Miller, Amos
Marshall, William
McKinney, D.
McKinnie, Amos
McFee, Mike
McArthur, Ira W.
McNutt, C. E.
McCauliff, J. D.
McAllister, Eliza J.
McFall, Mrs. Rhoda
McAllister, William
McNutt, J. W.
McNutt, J. E.
McNutt, Eliza
McAllister, Elmer
Milbourn, J. H.
Milbourn, Josiah
March, Watson
Miller, Frank
Miller, William
Meecham, Orlando
Moon, Mary
Mahan, Earl
Mast, Barry
Minnie, H. C.
Merritt, Geo.
Montgomery, Clifford
Miller, Peter
Miller, M. Irena
Markham, Maggie
Molton, Louisa
Menlock, Ira
Merritt, Chas.
Miller, Henry
Mowrole, George V.
Miller, Dennis
Miller, James C.
Marvin, H. D., estate of
Morse, Chas.
Miller, Uriah
Mendell, H. C.
Mitchell, Frank
Mendall, Edward H.
Morgan, Jo. K.
Maples, Harry
Narlcom, C. J.
Nichols, E. M.
Nash, Benj.
Newman, Mrs. E. S.
Newman, A.
Nelson, Geo.
Nash, Stephen
Nash, Mary
Osborn, E. C.
Osborn, F. A.
Owen, William B.
Osborn, A.
Otto, Will
Paris, James
Parker, Mrs. A.
Putnam, W. H.
Powers, Harry S.
Porter, Albert
Picket, Warren
Picket, C. E.
Payne, Geo.
Pennock, J.
Perry, James
Phillips, Wm. G.
Perine, Geo. C.
Pilmore, John
Powers, Calvin
Prestigen, Jane
Prescott, Lucinda
Priest, Atlantic
Plemons, Lucy J.
Porter, Lyman S.
Rose, Orrin
Rogers, Geo. S.
Rich, J. W.
Rogers, Mrs Garrett
Hornbeck, Marietta
Rogers, Will
Rogers, Sarah A.
Reeves, W. H.
Roberts, Isaac P.
Reynolds, Caroline L.
Rice, Miss Anna A.
Root, C. H.
Rew, Mrs. Rebecca
Sargent, Samuel
Rhead Bros.
Rogers, William N.
Reynolds, T. L.
Rhead, Benj.
Rier, E. A.
Rose, Nathan
Reynolds, Mrs. B. F.
Rayers, John
Reynolds, W. B.
Rew, James
Rogers, Prudence
Reynolds, Mrs. Almira
Rank Bella
Rank Reuben
Rice, F. W.
Ralph, J. W.
Reynolds, B. H.
Reynolds, I. N.
Rhead, H. O.
Raymer, George
Rogers, Mrs. O. K.
Russell, Luther
Ryan, Chas. H.
Ramsey, Mary
Reynolds, W. H.
Raymer, Chas. K.
Rossman, Gillet
Rank, Fred
Rogers, Mrs. Garret
Shaw, B. S.
Shaw, Sarah M.
Saofield, Jerry
Slocum, Lewo
Slayton, James R.
Speer, E. D.
Selbner, George
Swift, H. A.
Smith, Philetus
Stump, F. M.
Shaw Josephine
Salpier, L.
Smoke, Hiram
Smith, William
Smith, C. J.
Smith, J. W.
Shaw, J. C.
Sprinkle, M. B.
Slayton, Susan
Slocum, P. G. Mrs
Speer, E. M.
Snyder, Mrs. Alva
Stirling, W. F.
Spietz, Fred
Stirling, J. M., heirs
Stevenson, Chas
Sabin, Eugene
Sheldon, Allen & Co.
Soule, Hiram
Saird, E. M.
Smith, Nancy
Selby, J. C.
Smith, John G.
Strong, Hellen A.
Salsner, William
Stedwell, J. D.
Speer, Matilda
Stirling, Mrs. James
Shea, Edward
Shramp, Nelson
Stagt, A. D.
Sabin, Austin
Spleming, Hea
Spear, John H.
Strunk, Mrs. Sophiah
Stone, A. K.
Strawn, James
Saliaby, Mrs. Wm.
Stedwell, J. G.
Stowell, Mary A
Scott, J. P.
Sterling, Mrs. J. H.
Spencer, H. H.
Springer, O. H. P.
Strong, Mrs. D.
Seger, M.
Seger, Geo.
Seger, Cornelius
Sherwood, Mrs. M. O.
Springer, E. B.
Sprague, S. T.
Stevenson, John A.
Taylor, Martha.
Trelby, Mrs. Russell
Union Lumber Co.
Vrere, Mrs. John
Vaughan, H. O.
Vaughn, J. W.
Vaughan & Son
Vaughan, Wesley
VanGordon, John
Verac, Mae
Vell, G. W.
Walters, Mrs. Orrin
Webber, Samuel D.
Weight, V. N.
Waldron, John
Walter, Hiram
Wilkes, Chas. E.
Walters, Norman
Williams, S.
Williams, Thomas
Whitam, Geo. J.
Washburn, S. A.
White, L. T.
Wetherell, Willner
Whitaker, Geo. W.
Wood, Nelson, Estate of
Wells, Henry
Wilson, Mrs. Emma
Willis, L. B.
Whitaker, L. B.
Walters, Daniel
Willoughby, Samuel
Wright, Geo. L.
Wilkins, Francis G.
Wark, L. W.
Wack, Chas.
Waldo, Mrs. A. E.
Wells Geo.
Winn, Lee
Wells, I. C.
Wadsworth, M. M.
White, Mary E.
Wilson, Anna C.
Whitam, H. C.
Watson, Mary
Widger, Hannah.
Webster, H. F.
Walworth, O. J.
Weeks, Otis
Wood, James
Woodruff, Eugene
Wood, Jane
Walworth, Minnie A.
Washburn, John
Woodruff, Chas.
Webster, Geo. W.
Wilder, Geo.
Wood, F. M.
Wright, DeForest
Wells, Geo. Jr.
Wells, Geo. Sr.
Zimmerman, Mrs. S. A.

BENTON TOWNSHIP.

Austin, Fred 8	Charlotte
Bartholomew, Napoleon 23	"
Bartholomew, Levi 31	"
Blanchard, D. A. 31	"
Brooks, Alonzo 20	"
Brandeberry, D. A. 17	Potterville
Bare, Catherine 21	"
Baird, Otis 20	"
Barr, W. H. 24	"
Birkhead, J. F. 35	"
Brown, T. H. 15	"
Backus, Haight 13	"
Beard, Harrison 34 & 35	"
Brubaker, Jacob 24	"
Barker, H. B. 24	"
Burkhead, Herman 35	"
Beer, Amos 3	Grand Ledge
Cook, O. P. 2	"
Carman, J. F. 24	Potterville
Copley, Thomas 35	"
Carman, Richard 21	"
Carpenter, Frank 31	Charlotte
Casey, Peter 17	"
Cady, Fred A. 18	"
Cady, Geo. R. 19	"
Carpenter, Cyrus 19	"
Carpenter, H. F. 18	"
Challender, Mitchell 8	Charlotte
Cobb, William & Son 17	"
Cover, Anna 35	"
Cole, A. N. 30	"
Cole, N. J. 30	"
Casey, Thomas 31	"
Claflin, A. P. 31	"
Claflin, Mary E. 31	"
Claflin, A. F. 32	"
Cove, Anna 34	Potterville
DeCow, Hannah 6	"
Dutton, Mary A.	"
Briggs, Frank E. 29	"
Donaldson, James W. 21	"
Davidson, Susan 17	Potterville
Davidson, Harvey 17	"
Davies, H. L. 1	Grand Ledge
DeWitt, J. J. 1	Grand Ledge
Deveny, John 13	West Windsor
Eagle, Hannah 23	Potterville
Edman, O. P. 13	"
Edwards, James 1	"
Earl, William 3	Grand Ledge
French, Calvin 19	Charlotte
Fuller, D. P. 6	Roxana
Fenn, S. E. 3	Grand Ledge
Fry, Chas. 14	West Windsor
Ford, Chas. T. 13	Potterville
Fox, Herbert A. 12	"
Fox, Amos M. 24	"
Phaner, Chas. 35	"
Flanner, Andrew 23	"
Fast, Levi 36	"
French, Hiram C. 11	"
French, John H. 1	"
Garvey, P. H. 33	Charlotte
Godfrey, O. C. 29	"
Goodrich, Navy 10	"
Goodrich, Asa 17	"
Gregg, Walt 24	Potterville
Gudemark, William 16	"
Gudemark, Edward 16	"
Green, Cynthia 6	Grand Ledge
Green, Frank 6	"
Grant, C. H. 11	"
Griffith, William E. 1	"
Green, Geo. H. 2	"
Gregg, Robert 13	Potterville
Gregg, Mary 13	"
Hough, Geo. H. 29	Charlotte
Hoag, Clayton 4	"
Halliday, Lewis 10	"
Henry, T. K. 10	"
Hayes, H. P. 10	"
Haines, O. W. 10	"
Haines, F. T. 18	"
Hinton, H. C. 32	"
Hunn, David 43	"
Haltz, F. A. 31	"
Heller, C. W. 30	"
Hewlett, O. P. 28	"
Harmon, Henry 17	"
Harmon, Harry 16	"
Hanna, Silas W. 33	"
Malley, Enalon H. 34	Potterville
Malley, A. D. 27	"
Hummel, Michael 3	Grand Ledge
Holden, Munroe 5	"
Horner, H. W. 1 & 11	"
Hartell, Lewis 26	Potterville
Hartell, Frank 35	"
Hartell, Henry L. 23	"
Haynes, Perry F. 26	"
Haynes, H. M. 25	"
Higby, Lewis E. 1	"
Holstead, G. W. 10	"
Hulet, O. P. 29	"
Hough, Geo. P. 20	"
Hasen, A. L. 31	"
Johnson, Frank J. 10	"
Johnson, Fayette 3	"
Johnson, Geo. 34	"
Johnson, Elizabeth 34	"
Johnson, Luther 2	"
Johnson, Geo. 19	"
Jones, Chas. 11	"
Johnson, H. 16	"
Johnson, M. 10	"
Jackson, John H. 15	"
Kingman, William E. 20	Charlotte
Klingensmith, M. 33 & 34	Potterville
Kingman, J. G. 19	Charlotte
Kinsel, Geo. W. 33	"
Kiser, John L. 35 & 27 & 31	"
Kilmer, J. H. 30	"
Kilmer, Cyrus 31	"
Kilmer, Lillie 17	"

BENTON TOWNSHIP.

King, P. A. 19	Grand Ledge
King, Julia F. 1	Potterville
Laverty, Wheaton 12	"
Lett, Mason 22 & 21	"
Lett, Alice 20	"
Leake, Jennie B. 25	"
Lindsley, H. 13	"
Loehr, Jacob 15 & 16	Charlotte
Landers, Geo. W. 28	"
Lewis, William G. 21	"
Lester, Geo. 28	"
Lipsey, James H. 15	"
Lipsey, John 19	"
Lipsey, James Sr. 7	"
Lyon, Emery 7	Roxana
Leslie, John 1	Grand Ledge
Meyer, J. H. 10	Charlotte
Murray, V. B. 22	"
Mitchell, W. Z. 25	"
McIntyre, H. B. 28	"
McIntyre, Jerry 20	"
Miller, M. K. 31	"
McConnell, Myron 22	"
Mitchell, William 33	"
Munger, L. C. 26	"
Mann, William B. 21	"
Mason, Franklin 5	Roxana
Merrill, Fletcher 5	Grand Ledge
McCormick, Justin 3	"
McCormick, J. W. 3	"
Mulholland, Catherine 4	"
Metter, William 21	Potterville
Mosher, Isaac 12	"
Mitchell, W. A. 13 & 22	"
Munger, E. N. 14	"
Merrill, Frank 12	"
Merrill, Sarah 1	"
McDade, J. E. 26	"
Mauk, Geo. 27	"
Marshall, Sarah L. 15	"
Norton, J. B. 23	Charlotte
Norton, A. M. 21	Potterville
Newell, Nathan 29	"
Nixon, Jennie 1	Grand Ledge
Owen, John S. 8	Charlotte
Owen, Eva M. 8	"
Otis, W. D. 33	"
Otis, Asa L. 24	Potterville
Otis, Frank 12	"
Parker, John 14	"
Palmeter, Kara 30	"
Perry, L. & H.	"
Porter, Mrs. Alpheus 32	Charlotte
Perkins, J. P. 28	"
Pierce, C. V. 30	"
Parks, Samuel 1	Grand Ledge
Quantrell, W. H.	Potterville
Reed, Emily 34 & 35	Charlotte
Robinson, Ann 16	"
Rosier, Chas. 7	"
Rumery, S. 31	West Windsor
Rouse, Edward 14	Potterville
Root, John 36	"
Richards, A. B. 25	"
Rickard, P. W. 17	"
Berry, Bethuel 10	"
Rosengrin, H. E. 1	"
Roseman, H. F. 1	"
Rowe, E. 13	"
Shaver, Paschal 26 & 27	"
Shaver, O. C. 35	"
Shaver, Alfred 28	"
Shannon, A. R. 15	"
Squire, O. H. 10	"
Squire, R. 10	"
Stewart, Alfred 34	"
Sprout, Thomas 21	"
Shepherd, James 24	Charlotte
Straight, C. M. 33	"
Stebbins, William O. 33	"
Stebbins, Ransom 31	"
Shear, Marcus 30	"
Sly, Ella 7	Roxana
Tow, Thomas 7	Charlotte
Tow, Harry 6	"
Todd, J. W. 31	"
Tubbs, Geo. H. 29	"
Thomas, J. C. 9	"
Terrill, Lucy. 23	"
Thrall, H. M. 23	"
Thrall, David 27	"
Traeger, Paul 35	"
Tidd, Chas. 27	"
Toman, Geo. M. 13	"
Trumbull, Earl 17	Grand Ledge
Upright, Jacob 21	Charlotte
Upright, G. T. 16 & 22	"
Upright, T. D. 15 & 17	"
Underhill, A. B. 22	Charlotte
Underhill, Abraham 14	"
Vickery, Amanda 33	"
Vickery, Austin 32	"
Van Auken, C. F. 26	Potterville
Van Auken, Henry 26	"
VanZile, D. M. 7	"
Walker, Herbert A. 3	"
Wheaton, D. H. 4	"
Wheaton, A. M. 4	"
Welch, Irwin 26	"
Wigent, F. R. 15	"
Wigent, D. B. 15	"
Ward, David C. 11	"
Warren, Frank 14	Potterville
Weaver, Chas. 21	"
Weaver, Fred 16	"
Weaver, Seneca 9	"
Willis, Emma 27	"
Waltersdorph, W. J. 21	Charlotte
Whorley, Elmer 36	"
Woodworth, John 9	"
Wigent, G. M. 9	"
Woodworth, John 4	"
Woodworth, H. 4	"
Walsh, Thomas 20 & 21	"
Wise, David L. 6	"
Wilson, Mathew 2	Grand Ledge
Wright Joshua 7	"

VILLAGE OF POTTERVILLE.

Baran, F. H., laborer
Bennett, Asa, druggist
Bare, E. M., grocer
Beals, Frank O., laborer
Bragg, S. A., blacksmith
Bennett, Chas., laborer
Brown, T. L., laborer
Benton, Chas. R., laborer
Blodgett, C. A., carpenter
Bellows, Ransom J., salesman
Carpenter, H. L., miller
Conklin, Caroline, widow
Canfield, Elizabeth, widow
Croop, J. P., wagon maker
Dason, Jacob, laborer
Dyer, F. F., laborer
Darling, Alfred, railroad employee
Davis, David, laborer
Darling, Nelson, laborer
Darleen, B. F., carpenter
Dooley, Rebecca, widow
Emerson, D. W., jeweler
Elmore, Fanny B. widow
Edgett, William, laborer
Fairchild, C. H., retired
Frasier, J. D., retired laborer
Fordham, Benjamin, village marshal
Ferry, Geo., laborer
Gillett, Sam, retired
Gilbert, Geo., barber
Goodwin, A. P., post master
Hartwell, J. B., grist mill
Halsey, M. F., carpenter
Hickson, Harriet, widow
Haner J. C., laborer
Holmes, E. P., laborer
Hamilton, E. M., laborer, stone mason
Higbee, L. E. M. D.
Howe, W. H., shoemaker
Johnson, Henry, laborer
Johnson, Geo., farmer
Johnson, Elizabeth
Lock, Thomas, laborer
Lincoln, C. H., harness maker
Merritt, F. K., harness maker
Merritt, H. D., carpenter
Merritt, S. G., grocer
Marshall, Chas., painter
McBrien, Hugh, retired
Mathews, William blacksmith
Newell, Nathan, farmer
Needham, Alonzo, laborer
Odell, Mrs. E. W., widow
Olmstead, Simeon, laborer
Osterout, Isaac, laborer
Proctor, A. C.
Platt, Lafayette, laborer
Paine, Rowland, retired farmer
Potter, L.
Pittman, Amos, veteran soldier
Palmer, M. J., druggist
Parker, John
Proctor, O. W., painter
Quackerd, Belinda, widow
Rogers, Jay
Ross, I. T., building mover
Ray, Frank, laborer
Bowman, L. T., farmer
Redfield, Wm., blacksmith
Rowe, Edward, laborer
Smith, Alvin, laborer
Smith, C. L., retired carpenter
Smith, Frank S., barber
Spears, Thomas, farmer
Sykes, V. R., laborer
Sutliff, Theodore, blacksmith
Smalley, J. H., retired farmer
Stewart, Alvin, railroad laborer
Spears, Homer, farmer
Thompson, Ida
VanAuken, W. H., merchant
Van Zile, D. N., retired farmer
VanAuken, Henry, retired farmer
Winter, Jacob, engineer
Weinshaupt, Adam, drain commissioner
Williams, John, laborer
Welsman, A. M., laborer
Winters, Louisa, widow
Wilbur, Clara A., widow
Weaver, Seneca, farmer
Young, Mrs. E., widow

TOWNSHIP OF BELLEVUE.

Abbott, W. J.	Bellevue
Abby, S. M.	"
Adams, E.	"
Adams, A. R.	"
Adams, Nancy	Bellevue
Allen, H. M.	"
Amon, James L.	"
Amon, Mrs. J. M.	"
Andrews, Fred E. Sr. 20	"
Andrews, Fred E. Jr. 27	"
Armstrong, W. K.	"
Atherton, C. A. 17	"
Atherton, Tracey 3	"
Annis, Munson 20	"
Avery, Martin	"
Avery, C. G.	"
Babcock, Wm. H. 25	"
Bailey, Thos. 15	"
Beers, C. W.	"
Baggerly, Reuben 31	"
Bearcroft, F. F. 35	"
Berkimer, John 23	Olivet
Berry, E.	Bellevue
Barber, Maurice	"
Bond, Amelia	"
Bowen, Mrs. A. J.	"
Bisel, Frank 31	"
Billingsley, Harvey 35	"
Brown, O. L. 3	"
Brown, L. C. 8	"
Brown, W. A. 14	"
Butler, A. G.	"
Briggs, R. S.	"
Barton, Thos. 34	"
Billman, John 7	"
Campbell, Mrs. S. A.	"
Campbell, Fletcher 11	"
Cargo, Wm. Sr.	"
Cargo, Lucy	"
Carrol, Geo. 8	"
Cayton, Clara 35	"
Case, Augustus	"
Chismock, Jas.	"
Chittenden, Sarah	"
Chapman, Martha 23	"
Cole, S. W. 4	"
Colby, Dexter 36	Olivet
Colby, John 36	Bellevue
Cook, Caroline 36	Olivet
Cooper, Mrs. David 23	"
Coleman, Frank	Bellevue
Coleman, Isabelle	"
Cole, Fred	"
Cook, Wm.	"
Cook, W. H.	"
Cook, L. D. 27	"
Conklin, John 32	"
Cronk, W. F. 22	"
Cronk, N. H.	"
Cronk, Peter 3	Ainger
Crowell, David 16	Bellevue
Crawford, Mary 10	"
Cummings, Mrs. H. A. 5	"
Chapman, L. 5	"
Crowell, Geo.	"
Cushing, Jerome 22	"
Davidson, Jerry	"
Davis, Samuel	"
Davis, Theodore 65	"
Davey, Mrs. Wm.	"
Day, Curtis 1	Ainger
Day, A. J. 1	"
Day, L. J. 1	"
Dalton, John 8	Bellevue
Dickson, Geo. 29	"
DeBlamar, J. F. 27	"
DeMotte, Wm.	"
DePuy, Jacob 19	"
DePuy, Frank	"
DePuy, Warren	"
Denniston, W. S. 23	Olivet
DeBoe, Martha 35	Bellevue
Dimmick Lyman 29	Olivet
Dolph, Chas. 28	Bellevue
Dougherty, John	"
Dolph, W. W.	"
Denniston, Fred	"
Dyer, Chas.	"
Doolittle, Mitchell 6	"
Edwards, A.	"
Elmendorf, L. W.	"
Elmendorf, W. H.	"
Elmendorf, J. M.	"
Emery, Mrs. L.	"
Evans, John	"
Evans, S. B.	"
Evans, Levi 6	Ceylon
Fisher, Wm. 4	Bellevue
Follett, Willard 9	"
Folett, Anna 22	"
Flower, E. F. 11	"
Flower, A. D. 2	"
Fitzgerald, A. E.	"
Fitzgerald, Mrs. R.	"
Fitzgerald, D. W. 23	"
Fisher, A. C. 34	"
Frantz, Conrad 9	"
Foss, Wm. 31	"
Fruin, R. C. Jr. 31	"
Fruin, R. C. Sr. 30	"
Grubs, Jerome 34	"
Grayson, Sarah 24	"
Gage, Andie 26	"
Gibson, S. W. 24	"
German, Wm. 4	"
Gambel, Margaret 11	"
Gambel, Frank 11	"
Gambel, Chas. 11	Bellevue
Gayton, Wm. 10	"
Griffin, Morton 3	"
Greenman, W. M. 17	"
Gayton, W. H. 3	"
Garmen, Geo. 4	"
Garmen, Jacob 1	"
German, Geo. 4	Ceylon
Hamilton, Jas. 30	Bellevue
Hall, Nelson F.	"
Hunsicker, Henry	"
Holden, Geo. 22	"
Hess, H. S. 32	"
Howe, Wm. 27	"
Hughes, F. F. 33	"
Hunsicker, Lucy 33	"
Hodgeman, H. A. 15	"
Hall, J. B. 33	"
Hall, J. R. 33	"
Hull, H. D. 31	"
Hoyt, A. B. 33	"
Huggett, Wm. 23	"
Hoyt, C. M. 33	"
Higgins' Mary E. 23	"
Higgins, Cyrus 35	"
Harris, Lyman 20	"
Hamlin, Henry 20	"
Henlow, Mary 20	"
Hart, Harvey 20	"
Holland, Sarah	"
Hemenway, F. H.	"
Hine, Mrs. Geo.	"
Henion, Tim.	"
Hayes, Eugene	"
Hamlyn, Richard	"
Hall, Geo.	"
Hobbs, J. L. 27	"
Holt, W. H.	"
Holbrook, E. J.	"
Huggett, A. M.	"
Hibbard, Geo. H. 14	"
Hollenbeck, Moses 15	"
Hamilton, J. J. 3	"
Hoyt, Ezra 13	"
Hoyt, Daniel 12	Ainger
Hine, Geo. N. 14	Bellevue
Hall, Horatio 24	"
Hall, Henry R. 24	"
Hemenway, F. A. 28	"
Hamilton, J. W. 4	"
Hickman, Levi 3	"
Hart, Mrs. J. A. 28	Olivet
Hovey, M. A. 25	Bellevue
Hotchkiss, Lyman 6	Ceylon
Inman, Wm.	Bellevue
Inselman, J. P. 21	"
Inselman, Jacob 22	"
Johnson, Henry	"
Johnson, E. A. 10	"
Johnson, E. R.	"
Jenks, Royal 32	"
Johnston, Mrs. Jas.	"
Johnston, Wm.	"
Johnston, Mrs. Thos.	"
Johnston, Ella	"
Johnston, Clara	"
Johnston, L. G.	"
Johnston, Frank	"
Judd, Emma	"
Jenkins, John	"
Jackson, Mrs. Jas. 5	"
Jones, Victor 14	"
Jarvis, Philip 12	"
Jordon, Elliott 12	Ainger
Jennings, Oliver 22	Bellevue
Kingsbury, Mrs. E. M.	"
Kimberly, A. J.	"
Kimberly, C. B.	"
Kelley, Joel 21	"
Knapp, Edwin	"
Keith, Eugene	"
Kimberly, Orrin Jr	"
Kelley, Andrew 24	"
Lane, John	"
Latta, Wm.	"
Legge, Henry	"
Lucas, Mary	"
Larrison, D. W.	"
Larrison, Aurilla	"
Lane, Josephus 28	"
Lane, Harvey 15	"
Lane, James 15	"
Loomis, Mrs. Geo. 20	"
Lehner, Freeman 22	"
Luscomb, W. H. 21	"
Lennon, Hugh 8	"
Lennon, W. H.	"
Lehner, Wm. 18	"
Lindsley, S. D. 6	Ceylon
Lewis, Leonidas 16	Bellevue
Lankton, G. H. 19	"
Monroe, Alice	"
McGinn, Hugh	"
Mulvaney, Fred	"
Marshall, Frank	"
Meech, Wm. 22	"
Mason, Horace	"
Morgan, W. D.	"
Morgan, L. H.	"
Manary, John	"
McFall, S.	"
Monroe, Oscar	"
Martin, Mrs. Wm.	"
Mahoney, Jno.	Bellevue
Meech, Ellen	"
Madison, Jno.	"
Madison, Edward	"
Moon, Geo.	"
Moon, Wm.	"
Monroe, Clark M.	"
Monroe, G. W.	"
McManus, Frank 16	"
McManus, Wm. 16	"
McManus, Catharine 16	"
Moon, M. V. 17	"
McCarthy, Mary J. 15	"
Mulvaney, James 21	"
Meech, Mrs. James 21	"
Morton, Thos. P.	"
Mattison, Jas. 6	Ceylon
Murray, Bernard 1	Bellevue
Murray, Robert 4	"
Martins, John 7	"
Newton, Catherine	"
Newton, Warren	"
Nelson, Mrs. David 16	"
Niles, Geo. 27	Olivet
Newton, Ruth 25	Bellevue
Owen, A. J.	"
Osmun, Henry 27	"
Overholtzer, Hiram	"
Owen, E. F. 23	"
Owen, Chester 28	"
Pugh, Ruby L. 21	"
Palmer, H. H. 19	"
Pixley, W. E. 8	"
Pearsley, Kittie 31	"
Pfaff, Peter 16	"
Palmer, E. D. 35	Olivet
Plum, Geo. 21	"
Reed, Wm. 19	Bellevue
Robinson, Hiram	"
Robinson, Wm.	"
Robinson, Mrs. H. L.	"
Roberts, Johnathan 26	"
Rounds, Jennie	"
Rogers, Jane	"
Roberts, Mrs. Thos. 24	"
Ripley, Benj. 11	"
Rogers, M. D. 18	"
Rowe, Helmoth 14	"
Reynolds, E. M. 20	"
Roberts, James 20	"
Roberts, Johnathan 24	"
Roberts, Jacob 34	"
Spaulding, L. C.	"
Spaulding, Chas.	"
Spaulding, Jas. A. 21	"
Spaulding, E. R. 17	"
Spaulding, J. W.	"
Spaulding, Geo. G. 16	"
Seitler, Jacob 10	"
Stephens, Mrs. Ed.	"
Stephens, Richard	"
Sherbina, Mrs. F. F.	"
Sturge, Mrs. A.	"
Sawyer, A. J.	"
Sackett, Rachel	"
Spafford, L. E. Jr 10	"
Spafford, L. E. Sr.	"
Sackett, Frank	"
Stevens, Geo. P.	"
Stewart, Mrs. Ellen 24	"
Sherwood, Phydelia 24	"
Shatterman, Levi 16	"
Shumway, D. A. 28	"
Smith, Carrie 23	"
Smith, Junior 23	"
Sewell, Ira K. 12	Ainger
Spaulding, F. B. 8	Bellevue
Smith, Leonard 7	"
Shapley, Royal 7	"
Shapley, Philoma 7	"
Stevenson, L. 18	"
Stiles, Caroline 16	"
Sprague, Elmer 7	"
Spaulding, T. D. 14	"
Stevens, Frank 10	"
Stevens, Fred 10	"
Stevens, Lewis 10	"
Shatterly, David 10	"
Shatterly, Chas. 10	"
Shatterly, James 10	"
Secor, Chas. 11	"
Sharkey, Jas. 29	"
Seitler, John A. 3	"
Seitler, Melvin 9	"
Seitler, J. A. 9	"
Simon, Louis 4	"
Simon, Mrs. Nicholas 3	"
Simon, W. E. 4	"
Sargeant, Mary E. 8	"
Sackett, Alonzo 8	"
Shipp, Fred 26	"
Shipp, Wm. 26	"
Tobin, L.	"
Tobin, S. R.	"
Taylor, J. K.	"
Taylor, Martha	"
Trabert, William	"
Trowbridge, F. R.	"
Turner, C. A.	"
Turner, Elmer	"
Tracy, Mary	"
Tucker, S. H. 21	"
Tulley, E. 5	"

TOWNSHIP OF BELLEVUE.

Trialett, J. Est. 7	Bellevue
Treat, Bessie 17	"
Turner, A. K. 13	"
Thurston, J. M. 1	"
Thurston, Orvison 1	"
Thurston, Wm. 1	"
Thornton, A. B. 1.	"
Tillotson, Henry 2	"
Toole, Lawrence Mrs. 30	"
Van Valkenburg, Mary	"
Van Valkenburg, Mrs. A.	"
Vlemaster, Henry 3	"
Vlemaster, John 3	"
Van Scotter, S. 27	"
Van Scotter, E. 25	"
Van Sickle, Orrin 11	"
Vedder, Levi 19	"
Wells, A. S. 27	"
Willis, Geo. 34	"
Willis, Ida 31	"
Willis, Fred 34	"
Wood, Ida 28	"
Wiley, Milo	"
Wilson, Marion	"
Woodruff, Mrs. S. A.	"
Wilson, Frank	"
West, Willard 29	"
Wilson, Griffin	"
Walkinshaw, James	"
Way, Chas.	"
Walker, G. W.	"
Winslow, Frank	"
Wood, Ellen J.	"
Wilson, Harriett 30	"
Williams, Elijah 23	"
Wiles, Henry 7	"
Wight, Mrs. M. A. 18	"
Young, Claude 11	"
Young, Harrison 11	"
York, Chas. H.	"
York, John	"
York, Judson	"
York, Chas.	"
Young, Chas.	"
Young, S. H.	"

BROOKFIELD TOWNSHIP

Atwood, Aldro. Est. 2	Charlotte
Aulls, Rosalthe 4	"
Annis, William 27	Brookfield
Ayer, John, Est. 25	"
Annee, James 21	"
Anson, Frank 20	"
Andrews, J. G. 31	Partello
Basler, Martha A. 9	Charlotte
Barkstall, J. H. 7	"
Blood, F. A. 5	"
Ball, Marion 17	"
Bryan, Horace 7	"
Braman, John 8	"
Blanchard, C. 25	"
Buskirk, James 25	Charlesworth
Briminstool, Jas. J. 23	"
Brimlinstool, Chas. M. 15	"
Brimlinstool, James M. 24	"
Brimlinstool, Mary A. 24	"
Bingham, A. L. 31 & 33	Charlotte
Baum, Daniel 21 & 25	Charlesworth
Barrett, N. 28	Brookfield
Brangwin, William 20	"
Beck, Robert P. 31	"
Bickford, Egbert 6	"
Bryan, D. W. 28	"
Baxter, John 17	"
Ball, W. A. 27	"
Buckman, Alice M. 27	Eaton Rapids
Bateman, Leonard 12	"
Brimlinstool, Minerva 33	"
Bacon, Adelbert 2	"
Brainard, Zerah 34	"
Bly, William H. 13	"
Bandt, Christian 1	"
Buckland, G. W. 23	"
Bly, Samuel 13	"
Bly, Sylvia 13-19-20-27 & 35	"
Ball, Dina 19	Brookfield
Bundy, Adelbert 33	Eaton Rapids
Cornell, Thos. 4	Charlotte
Carpenter, John 1	Eaton Rapids
Crouse, John 26	Brookfield
Cooper, Celia 34	"
Chaney, Jared 27	"
Cross, Jas. A. 22	"
Crandall, Frank 30	"
Coats, Elihu 27	"
Cutler, L. A. 28	Charlesworth
Carrier, H. L. 33	Brookfield
Croop, Phillip 7 & 8	Charlotte
Croop, G. S. 7-8 & 18	"
Dickinson, Carrie, Est. 5	"
Dewey, F. E. 30 & 34	Brookfield
Devonshire, Henry 28 & 31	"
Dingman, F. L. 24	"
Demier, J. B. 1	Eaton Rapids
Demier, A. D., Est. 2	"
Decker, Levi 31	Springport
Edick, Alex 6	Charlotte
Edick, F. A. 6 & 7	"
Edick, Peter 5 & 9	"
Edick, Lewis 8	"
Edick, William 8	"
Eschenbacher, Louis 1	"
Ellison, Samuel 11	Brookfield
Eaton, William 23	"
Estelle, John G. 27	"
Estelle, Edwin 21	"
Estelle, Henry 25	"
Edwards, Albert 15	"
Englehart, Adam 14 & 28	Eaton Rapids
Epley, H. K. 19	"
Fast, J. H. 18	Charlotte
Fast, G. A. 18	"
Force, Geo. 1	"
Frank, John, heirs 6	"
Frank, Mrs. S. M. 6	"
Flatter, A. H., Est. 8	Brookfield
Frove, Lloyd 27	"
Fordham, Clarence 20	"
Finch, Asahel 20	"
Ferris, Chas. 32	"
Finch, A. C. 17	"
Fox, Edwin 10	Eaton Rapids
Fox, Martin 3 & 10	"
Faulkner, John 31	Partello
Faulkner, Mervin 19-30 & 31	"
Favorite, Thomas 21	Charlesworth
Favorite, John 22	"
Favorite, Geo. 21	"
Fuller, Gilbert 31	Springport
Fancher, Mary Jane 29 & 32	Brookfield
Gelabert, Mrs. 5	Charlotte
Goshorn, L. 28	Brookfield
Gibson, Silas 12	Eaton Rapids
Gilman, Harling H. 11	"
Goodrich, Dessom 11	"
Gilders, Jas. 26 & 27	Brookfield
Green, E. H. 33 & 34	"
Hall, Chas. 20	Charlotte
Hipp, Fred 5	"
Hull, John 4 & 7	"
Holbrook, Aruna 10 & 11	"
Holbrook, Ernest 11	"
Harrow, Frank 6	"
Higby P. M., Est. 5 & 6	"
Hart, Jesse Jr. 6	"
Hoey, John 1	Brookfield
Hahn, D. A. 21	"
Henderson, C. T. 16	"
Hawley, T. B. 23	"
Hawley, Henry 17	"
Hall, Geo. 20 & 21	"
Hall, Clarence 14	"
Hall, James 20 & 30	"
Hotchkiss, G. E. 21	"
Humphrey, Ira 15	"
Harris, Chas. 14-15 & 21	Eaton Rapids
Heath, Chas. P. 12	"
Henry, Lydia 11	"
Huey, Elizabeth 5	"
Hutchings, F. L. 30	"
Jones, Theodore 29	Brookfield
Johnson, Ira 17	"
Kay, John D. 16	"
Kellogg, D. C. 21	"
Kock, E. R. 28	"
Krebs, A. 10	"
Krebs, Eli 8 & 17	"
Krebs, Levi 23	"
Krebs, Henry 23	"
Kessler, Marion 21	"
Leonard, Geo. 9	Eaton Rapids
Leonard, W. S. 9	"
Lown, Sebastian 15	Charlotte
Logan, D. A. 28	Brookfield
Lounsberry, Allen 30	"
Lounsberry, Judson 20 & 21	"
Livingston, Henry 10	"
Leightner, Samuel 26	"
Mahan, Henry 14 & 15	Charlotte
McDonald, Henry 21	Brookfield
Miller, L. C. 14 & 20	"
McClintick, Warren 17	"
Miller, Merton 23	"
Meinz, Sarah A. 16	"
McGowan, James 21	"
Myers, Joseph 33	"
Mitchell, Rolland 24	Eaton Rapids
Magron, Chas. 2	"
McManus, Alfred 1	"
Moses, John 22	Duck Lake
McCutcheon, J. P. L. 9 10-16	Brookfield
Newton W. M. 21	Brookfield
Newlon, W. S. 17	"
Number, John 28	"
Nevins, L. P. 31	Springport
Nicolls, Julius 33	"
Owen, G. W. 30	Charlotte
Olney, Richard 24	Brookfield
Olger, Philinda 29	"
Parr, William, Est.	Charlotte
Pierce, Mrs. Kate 12	"
Preden, J. C. 8	"
Peck, William N. 30	Charlesworth
Payne, Emery 20	Brookfield
Perry, H. R. 18	"
Perry, B. A. 21	"
Perry, S. J. 17	"
Pike, Chas. 21	"
Potts, William 21	"
Peters, J. M. & Son 33	"
Phillips, Walter 29	"
Phillips, Oda 30 & 1	"
Phillips, Wilber 29	"
Powers, William 30	"
Powers, Joseph 29	"
Pettibone, H. Est. 5	Brookfield
Pierce, C. E. 31	Springport
Peacock, Chas., Est. 30	"
Pratt, Lewis 27	"
Post, B. W. 13	Eaton Rapids
Peck, Abbie, 24	"
Peck, F. M. 24	"
Post, Josephus 1	"
Prescott, Joseph 1	"
Prescott, Harriett 1	"
Peck, Arthur 13	"
Post, B. W. 18	"
Post, Henry 14	"
Post, W. H. 12	"
Rater, Alvin 7	Charlotte
Rogers, Mrs. W. A. 8	"
Rush, Alonzo 18	"
Ripley, David 8	Brookfield
Riddle, C. H. 30	"
Rose, Rufus 33	"
Reynolds, Frank 31	"
Rochester, Wilbur 28	"
Reymer, Geo. 24	Eaton Rapids
Raymond, Mary 14	"
Raynor, Albert 23	"
Rochester, Daniel 14	Charlesworth
Rowe, Chas. 24 & 32	Springport
Rose, Elizabeth 3	Eaton Rapids
Rudesill, Clark 31	Partello
Rudesill, J. B. 31	"
Starkweather, Geo. A. 6	Charlotte
Starkweather, Geo. M. 6	"
Shaver, Homer 6	"
Spotts, Geo. 5	"
Spicer, Fred A. 5	"
Spicer, Frank A. 6	"
Spicer, W. A. 6	"
Spicer, Alfred 6	"
Sackett, A. L. 5 & 9	"
Spotts, Ed. 9	"
Shaver, Chauncey 9	Brookfield
Sherman, Fred. 9	"
Stead, Catherine 31	"
Sowle, Anna 34	"
Sowle, Willis 34	"
Smalley, Mary E. 20	"
Seybolt, Mrs. 20	"
Spears, Jas. 30	"
Stuart, Myron 14-15 & 23	"
Stevens, Irving 9 & 15	"
Smith, Thos. E. 16 & 21	"
Sowle, John B. Estate of 27	"
Starks, T. H. 31	Charlotte
Stevens, Asa 25	Brookfield
Scott, Eugene 31	"
Seefried, Geo. 28	"
Southard, Ed 19	"
Saulsbury, Mrs. 19	"
Sheldon, Elmer 29	"
Stuart, Sarah Jane 23	"
Seybrist, Era 18	"
Snyder, David 1 & 8	Eaton Rapids
Swan, W. W. 11-13 & 14	"
Swan, Adelbert 14	"
Swan, Morgan 11	"
Snyder, Henry, F. 2	"
Steffey, Henry 2	"
Shicker, George 12	"
Seevers, Geo. 13	"
Shicer, Ezra 23	"
Shiderman, Geo. 13 & 14	"
Short, Thos. 1	"
Sowles, Chas. 25	Charlesworth
Sutton, David 27	"
Steele, John 21	"
Sherman, Geo. 32	Springport
Sholtz, William 30	"
Starks, Dorsten 20	Duck Lake
Schrontz, Chas. 10	Brookfield
Thuma, George 30	Eaton Rapids
Thuma, Simon 23	"
Thomas, Jay 13	"
Thomas, Nancy 11	"
Tolbert, William 15	"
Taylor, Geo. 27	Brookfield
Taylor, Thomas 23	"
Taylor, John W. 22	"
Taylor, David 8 & 15	"
Taylor, W. H. 27	"
Tice, J. D. 8	"
Tolbert, Michael 18 & 19	"
Topliff, Elijah 15	"
Troutner, M. S. 25	"
Tulip, Geo. 4	"
Tulip, Ed 9	"
Unterkircher, C. 12	Eaton Rapids
Umbarger, Jas. 9 & 5	"
Voorman, Newton 21	Brookfield
Vanande, Catharine 16	"
Vanande, M. E. 16	"
Voxbury, Burt	"
Vanauken, Jas. 11	"
Williams, A. M. 6	Charlotte
Wells, S. P. 18	"
Williams, Jesse 21	Brookfield
Williams, Levi 27	"
Wixson, Albert 27	"
Wixson, Alice 34	"
Wixson, Willis 34	"
Walling, David 17	"
Wittman, J. J. 43	"
Walker, Julia 30	"
Webster, Marie 28	"
Webber, Joseph 16	Brookfield
Williams, Mary A. 20	"
Watkins, J. 30	"
Westgate, Frank, 25	"
Whitaker, G. W. 21	"
Williams, Latson 21	"
Williams, Samuel Park 23	"
Wilder, G. W. 32	"
Wilcox, Wesley J. 25	"
Williams, Mrs. D. T. 20 & 29	"
Warren, John 24	"
Warren, John T. 26	"
Whittum, Edgar 10	"
Whittum, C. U. 3 & 10	"
Wohr Reinhart 1	"
Walworth, John 10-11-14 & 15	"
Walling, John 32	"
Walker, William 35	Springport
Wilber, H. 30	"
Wagoner, Peter 21	"
Yeager, George 11	Eaton Rapids
Yoshner, Milo 21	Brookfield
Zeigler, Christ 30	"

CARMEL TOWNSHIP.

Allen, Mrs. Estelle 20	Charlotte
Amspacher, Israel 6 & 7	"
Amspacher, C. 7	"
Amspacher, E. 7	"
Amspacher, Mrs. L. 7	"
Ash, Isaac 5	"
Antell, Henry 32	Alnger
Benham, Albert 31	"
Bowsher, Edward 10	Carlisle
Bisel, Chas. 8	Charlotte
Ballanger, G. 16	"
Brighty, Gregory 9	"
Bedell, Emerson 24	"
Bolock, John 1	"
Blackmore, Mrs. James 21	"
Baker, Mrs. B. H. 4	"
Duhn, B. H. 5-7	"
Bowers, Jas. 20	"
Hodgetts, Geo. 6	"
Bowers, Palmer 4	"
Backus, M. 36	"
Baker, Francis 8	"
Baker, T. H. 5 & 9	"
Black, G. W. 5	"
Crabtree, Mrs. Ellen 17	"
Cushing, Frank 27	"
Cortright, H. 21	"
Chase, Archibald 33	"
Cameron, Frank 3	"
Clements, Samuel 5 & 6	"
Crabtree, Seth 17	"
Case, C. H. 21	"
Case, W. 21	"
Cooper, A. B. 21	"
Cooper, Harris 28	"
Cooper, J. B. 29	"
Cooper, M. W. 24	"
Coryee, James, 32	Alnger
Cooper, G. H. 15	Carlisle
Corsett, G. P. 30	"
Cole, H. D. 31	Alnger
Dawson, Chas. 10	Charlotte
Dillon, Amos 20	"
Demoing, E. 15	"
Davis, Russell, 9-10 & 11	"
Dillon, Thomas 27	"
Davis, Homer 29	"
Davis, Warren 16	"
Davis, E. W. 16	"
Dillon, L. T. 23	"
Dulball, Geo. 35	"
Exander, Alma 36	"
Ells, J. H. 29	"
Foster, Warren 10	"
Foster, Mrs. I. J. 17	"
Foster, Ira 16	"
Frace, Isaac 29	"
Frace, Geo. 5	"
Fowle, Martin 29	"
French, Edmond 29	"
Fickes, Mrs. M. 17	"
Ferguson, Stephen 4	"
Fortney, John 5	"
Fanney, David 17	"
Frace, Warren 19	Carlisle
Foote, Alton 32	Alnger
Griffin, Fred 17	Charlotte
Griffin, John 33	"
Griffith, J. Q. 33	"
Green, W. L. 1	"
Griest, Bert 10	"
Grier, David 19	"
Griffin, Albert 30	Carlisle
Gage, G. H. 30	Charlotte
Gussey, Chas. 4	"
Griffin, S. R. 26	"
Granger, T. 6	"
Gregg, Mrs. John 14	"
Griest, C. H. 11	"
Halsey, Chas. 36	"
Hutchings, Gilson 13	"
Halldaus, A. 17	"
Horn, Peter 27	"
Horn, J. P. 27	"
Howe, David 26	"
Holmington, Martha 18	"
Harmon, C. 9	"
Harmon, William 2	"
Halsey, Chas. 21 & 33	"
Hodge, Chester 9	Charlotte
Hoot, Elizabeth 21	"
Hilburn, J. N. 2 & 3	"
Harlow, C. R. 1	"
Hendricks, J. D. 31	Alnger
Hobler, William 20	Charlotte
Huber, Daniel 2 & 13	"
Jackson, Henry 13	"
Knapp, Delos 25	"
Knapp, Byron 21	"
Kelley, Lewis 7	"
Kelly, Mrs. A. 7	"
Kent, Johnson 23	"
Krusen, Geo. 25	"
King, Frank 33	"
Krebbs, Alex 36	"
Krusen, Mrs. Isaac 27	"
Kent, E. 33	"
Krusen, Enoch 33	"
Klein, Gren 1	"
Love, C. M. 8	"
Lane, J. N. 11	"
Lentz, August 10	"
Lake, J. P. 2	"
Laurence, Estate 1	"
Lawhead, Mrs. J. 20	"
Love, D. H. 29	"
Love, H. 29	"
Mitcwell, Frank, 27	"
Mitcwell, Levi 27	"
Mount, Will 21	"
Mount, William 8	"
Mygrants, S. A. 23	"
McClintock, Lucius 27	"
McWethy, Warren 29	"
Miller, D. S. 9 & 10	"
Miller, Jacob 9	"
Miller, Mrs. Jacob 10	"
McCreery, Mrs. 19	"
McCreery, Geo. 19	Carlisle
Morris Eliza 19	"
McCreery, Elvin 19	Charlotte
McCreery, Daniel 19	"
Mellon, Sarah 19	"
McCrumb, Chas. 36	"
Mulholland, James 11	"
McClintock, Fred 16	"
Maurer Richard 21	"
McBride, Samuel 34	"
Mulholland, Mrs. F. M. 22	"
Miller, Orrin 36	"
Mount, J. D. 9	"
Mason, P. M. 17	"
Merritt, W. W. 23	"
Mason, P. M. 5	"
McLaughlin, Mrs. J. 8	"
Maurer, Jos. 27	"
Mosey, Ed 10	"
McConnell, C. H. 30	Carlisle
Newcomber, Lewis 16	Charlotte
Norton, Phebe, Estate of 29	"
Nichols, Geo. 23	"
Nichols, A. A. 8	"
Peters, C. D. 30	"
Pollard, William H. 16	"
Palmer, Philo 35	"
Pierce, T. C. 1	"
Pollard, John 16	"
Ramsey, David 8	"
Rounds, Daniel 18	Carlisle
Reynolds, C. L. 14	Charlotte
Reed, Watson 20	"
Ray, Jos. 11	"
Roberts, D. F. 5	"
Rogers, Marcus L. 4	"
Raymond, Kingley 2	"
Randall, Susanna 18	"
Ripley, M. R. 36	"
Rich, John 11	"
Raymond, Henry 11	"
Stratton, Albert 10	"
Snyder, Kate 8	"
Sykes, C. 17	"
Stall, James 31	"
Sprague, Mrs. E. 10	"
Staley, L. S. 29	"
Shaver, John 6	"
Sherman, J. D. 22 & 15	"
Stacey, L. S. 20	"
Schauneckenburger J. 21	"
Schaueckenberger, Fred 23	"
Sline, Mrs. Daniel 4	"
Sternberg, Ellen 25	"
Sherman, J. H. 21	"
St. Clair, J. 14	"
Sprinkle, Amos 8	"
Silverthorn, Geo. 32	"
Smith, E. B. 6	"
Stall, D. J. 33	"
Snyder, M. 5	"
Snyder, J. H. 2	"
See, Eldridge 21	"
Smith, P. P. 35	"
Shaull, David 28	"
Saunder, William B. 33	"
Nixon, Herbert 33	Carlisle
Sikes, D. F. 30	Charlotte
Sikes, C. H. 31	"
Sikes, J. J. 30	Carlisle
Stewart, D. H. 31	Alnger
Tower, Knowles 15 & 16	Charlotte
Thornton, Gilbert 30	"
Thornton, Oscar 23	"

CARMEL TOWNSHIP.

Thompson, Frank 33 & 35	Charlotte
Tanner, Frank 7	"
Todd, Will 5	"
Terrill, Fred 24 & 25	"
Terrill, Judson 23	"
Terrill, Frank 25	"
Todd, Sylvanus 21	"
Tanner, John M. 19	Carlisle
Tanner, C. H. 18	"
Vanderhoof, Warren 16	Charlotte
Whitney, Anson 28	"
Wilkins, John 1	"
Wing, W. E. 1	"
Wood, L. W. 19	"
Wilson, Daniel 23	"
Whitcomb, Frank 24	"
Whitcomb, William 20	"
Walters, Levi 25	"
Williams, Z. 1	"
Wile, Daniel 10	"
Wade, G. H. 16	"
Williams, Geo. 31	[illegible]
Will, Wm. 19	Carlisle
Young, L. W. 10 & 15	Charlotte
Young, G. M. 11	"

CHESTER TOWNSHIP.

Andrews, C. T. 21	Gresham
Andrews, Ida 19	Chester
Ames, J. H., Est., 18-19 & 20	"
Ames, Frank 28	"
Ames, James K. 17-18	"
Ames, Wm. Sr. 20-29	"
Ames, Wm. Jr. 18	"
Arnold, Frank, 20	"
Ash, Austin 25	"
Barr, Clarence 21	Charlotte
Blair, Wm. 21	"
Bennett, M. M. 33	"
Bennett, Lucy 5	"
Barnley, Wilbur 15	Gresham
Barnley, Elizabeth 20	"
Bissell, Amanda 14	Charlotte
Boyer, Edwin 10 & 11	Roxand
Boyer, Joseph 10	"
Bowman, Frank 21	"
Bowman, Gilbert 1	"
Bosworth, Kelley 6	Bismarck
Bosworth, Myron 6	"
Bosworth, Edwin 5	"
Beckman, B. F. 4	Vermontville
Braden, Jasper 13	Charlotte
Bradley, Noble 14 & 24	Gresham
Barker, Gideon 14	Chester
Baker, L. 27	"
Blodgett, Anson 31	Charlotte
Blodgett, LeRoy 34	"
Bowman, Mrs. E. J. 34	"
Bolock, Jno. J. 34	"
Bolock, Wm.	"
Betts, Adon 35	Chester
Bissell, Cecilel 31	"
Bissell, Hiram 31	"
Bowers, Geo. 33	"
Baerr, Edward 33	"
Bates, James 36	"
Bottomly, James 30	Vermontville
Bell, Henry 30	Chester
Boyles, Wm. 31	Charlotte
Brinkham, Wm. C. 32	Chester
Barbyte, E. V. 10	Gresham
Barbyte, Chas. 5	"
Barnley, Chester 9	Gresham
Case, J. W. 30	Charlotte
Case, Mrs. Lemuel 33	Chester
Carter, Geo. W. 29	Charlotte
Carpenter, Cyrus 23	"
Carpenter, Francis 13	"
Cole, J. A. 21	Chester
Collins, L. R. 31	"
Clements, Wm. 29	"
Copples, Chas. 27	Vermontville
Crocker, Wm. 32	Chester
Clark, Nathan 24	"
Clapper, Orvis 20	"
Curry, Mrs. Clarissa 31	Charlotte
Colvin, John 9	Gresham
Davidson, Henry 4	"
Davis, Chas. 1	Roxand
Dean, Chas. W. 6	Vermontville
Doland, D. E. 9	Roxand
Dorman, Julius K. 24	Chester
Earl, Rufus 12	Roxand
Earl, Reuben 11	"
Eason, Geo. Sr. 23	Chester
Ellison, Jas. E. 28	"
Emry, Elim 26	Charlotte
Emry, Orvin 24	"
Farrah, James E. 30	Chester
Farrah, Chas.	Gresham
Ferguson, J. J. 1	Roxand
Fillcock, Levi 16	"
Field, Jas. 9	"
Fisher, Otis 3	Roxand
Fletcher, Geo. 4	Gresham
Fleury, Alvin 13	Charlotte
Foote, Mrs. T. M. 25	"
Foote, Frank 26	"
Francis, Lewis 5	Bismarck
Ford, Albert 35	Charlotte
Freed, H. H. 10	Gresham
Fuller, Wm. P. 11	Roxand
Fullerton, Alex 35	Chester
Gallagher, Catherine 34	Charlotte
Garn, Jessie 14	"
Geers, Bert 10	Roxand
Geer, Asbel 1	"
Gibbons, Edward 36	Charlotte
Goodenough, Mrs. T. 4	Bismarck
Grant, Mrs. S. 20	Chester
Grant, Dewitt C.	"
Grant, Thomas 21	"
Green, W. C. 14-22-23	Charlotte
Griest, Wheaton 33	"
Haigh, Daniel 6	Gresham
Hale, Wm. B. 1	Roxand
Hale, T. A. 4	"
Hale, Milton K. 35	Charlotte
Hampton, Daniel 15	Gresham
Hampton, Wm. 6	Roxand
Hampton, Geo. 6	"
Hampton, Chas. 1	"
Hampton, Earl 15	Charlotte
Hampton, Henry 29	Gresham
Harrington, Frank 18	Roxand
Harmon, S. W. 16	Gresham
Harmon, I. L. 9	"
Henry, T. E. 21	"
Hicks, Wm. 10	"
Hildreth, Hiram O. 22	Charlotte
Hildreth, Jennie 17	Gresham
Howe, Lewis 28	Chester
Hubbard, Robt. 27-34	Chester
Hutchins, Lee 35	Charlotte
Johnson, Mrs. E. A.	"
Jordan, Frank S. 21	Chester
Kirk, Jefferson 33	"
King, Paul 9	Roxand
King, J.	"
Klin, Geo. 14	Chester
Knapp, Silas 25	Charlotte
Knapp, Geo. 25	"
Lamie, Geo. W. 27	Chester
Lamie, Mrs. John 27	"
Lamont, Seth A. 28	"
Lipsey, James 13	Charlotte
Loomis, O. C. 4	Gresham
Lords, James A. 30	Chester
Lords, E. D. 32	"
Lords, Danielette 27	"
Love, H. T. 1	Roxand
Love, S. A. 1	"
Loveland, David A. 35	Charlotte
Loveland, E. E. 31	Chester
Lyon, Mrs. J. M. 23	Chester
Markham, O. J. 20	Charlotte
Mason, Ira 18	"
Mason, S. 1	"
Mason, Frank 1	"
Mason, Wilbur 12	"
Martin, E. H. 14	"
McDonald, Frank	"
McCormick, Jeremiah 24-25	"
McGonnley, Wm. 15	Gresham
McLaughlin, D. 31	Chester
McGargar, Chas. 4	Roxand
Merritt, Solomon 10	"
Messner, Mrs. Mary 24	Chester
Miller, W. D. 18	Gresham
Miller, Jas. 33	Chester
Mitchell, David M. 1 & 2	Charlotte
Mitchell, Chas. 1 & 12	Roxand
Moore, Annie 21	Gresham
Moore, Adelbert 21	"
Moore, Arthur 21	"
Moore, Allison 31	Charlotte
Moore, Philip P. 24	"
Mosher, Chas. M. 28	Chester
Mosher, Lee 26	"
Moyer, Hayden 3	Roxand
Moyer, W. I. 3	"
Moyer, Henry A. 3	Charlotte
Murray, James 32	Chester
Nash, Derahan 35	Charlotte
Newcomer, John W. 26	"
Neville, Wm. 21	Chester
Nichols, Robt. 33	"
Otis, Mark D. 14-20	"
Otis, Eugene 14	"
Olney, Geo. 29	Carlisle
Oumpocher, Ira 31	Chester
Oumpocher, Emanuel	"
Okinney, Jas. 17	"
Okinney, Andrew 18	"
Parr, Herbert 11	Charlotte
Parr, Ernest 11	"
Perkins, Mrs. M. A. 25	"
Perkins, J. H. 4	Roxand
Pierce, Thos. 19	Vermontville
Pierce, A. J. 20	Chester
Pitcher, Edmund 13	Charlotte
Phelps, Daniel 11	"
Phelps, Wm. 14	"
Phillips, Oscar 21	Gresham
Phillips, Mary 29	"
Phillips, Harlow 7	"
Pothemus, Mrs. B. 13	Charlotte
Pope, Elijah 33	Chester
Prindle, H. S. 24	"
Prindle, Uri 14	Charlotte
Prindle, Parintha 14	"
Pugh, Sheldon 34	Chester
Pugh, Alice J. 33	"
Quantrell, Chas. H. 35	"
Ransom, Geo. 35	Charlotte
Rasey, Armon 33	Chester
Rich, J. A. 13	Gresham
Rix, Geo. & Co. 1	Roxand
Robison, Seth 24	Charlotte
Rogers, Samuel 6	Gresham
Rogers, Ivert 7	Vermontville
Rogers, Clark 12	Gresham
Roll, John 4	Roxand
Root, Geo. 9	Gresham
Russell, Robt. 30	Chester
Sackett, Edwin S. 27	Gresham
Sackett, J. A. 15	Charlotte
Sanlow, J. L.	Roxand
Satterly, J. H. 17	Gresham
Satterly, E. 9	"
Satterly, Byron 9	"
Satterly, Frank 6	"
Satterly, Newman 10	"
Scott, Anson 27	Charlotte
Scott, Edwin 18	Vermontville
Scott, Alfred 18	"
Scott, L. R. 4	Gresham
Scott, Chas. 4	"
Schrader, J. C. 28	Charlotte
Sibbrell, Geo. 25	"
Simpson, Leslie 11	Roxand
Simpson, Nelson 14	Gresham
Shaw, John 27	"
Shaw, Chas. Sr. 32	Chester
Shaver, Jos. W. 34	"
Schermer, Wm. 26	Charlotte
Summers, Jacob 33	Chester
Smart, Asa 17	Gresham
Smart, Chas. Sr. 10	"
Snyder, Jno. Sr. 25	Charlotte
Spafford, Arthur 33	Chester
Sprague, Sanders 19	Vermontville
Stall, James D. 33	Chester
Stall, C. W. 15	Gresham
Stall, Rachel 17	"
Stoffer, Wm. 18	Chester
Strickland, L. A.	Gresham
Tanner, W. A. 33	Charlotte
Thompson, Thos. 20	Gresham
Tow, Jno. 31	Charlotte
Tow, Jas. 31	"
Trescott, Mrs. C. 19	Gresham
Treadwell, Eli 31	Charlotte
Turner, O. A. 28	"
Turner, Byron 28	"
Turner, Myron 21	"
Turner, Mrs. Chas. 27	Gresham
Uhl, James 14	"
Uhl, Lora 24	"
Uhl, Wm. 12	"
Vantassell, L. 20	Charlotte
Vincent, Wm. 8	Gresham
Waddell, Preston 15	"
Waddell, Mrs. E. 34	Charlotte
Ward, G. B. 15	Lake Odessa
Warner, Asa 19	Vermontville
Waltersdorf, D. J. 28	Charlotte
Wait, B. L. 15	Gresham
Wait, Adelbert 5	"
Walker, Mrs. Ellie 25	Charlotte
Webster, Mrs. A. E. 24	"
Wells, Daniel 30	Chester
Wells, Harvey 21	"
West, Frank 21	Gresham
West, Wilbur 16	"
Wheaton, Lemuel 22 & 27	Charlotte
Wheaton, Wesley 18	Roxand
Winston, Frank 20	Charlotte
Wheeler, Anna M. 34	"
Wheeler, D. D. 27	"
Waite, J. J. 31	Chester
Williams, James A. 33	Charlotte
Williams, Morris 31	Gresham
Williams, Bert 21	"
Williams, Levi 35 & 22	"
Williams, Geo. A. 21 & 22	Charlotte
Wilson, Oscar 33	"
Wright, Jas. 32	Chester
Wyble, Fred 28	"
Wyble, Frank 22	"
Wyble, Chas. 23	"
Young, M. F. 30	"

DELTA TOWNSHIP.

Atfatter, J. A. 34	Millett
Atwood, Chas. 15	Lansing
Atwood, W. L. 16	"
Allen, Alice 16	"
Alexander, A. F. 2	Delta
Baldrick, William 3	"
Banson, Christian 11	"
Bird, Hiram 21	Grand Ledge
Beech, M. F. 8	"
Briggs, James 18	Lansing
Billings, W. B. 16	Grand Ledge
Boehler, John 16	"
Bristol, Jerome 3	Delta
Brader, Henry	Millett
Bird, M. H. 35	"
Bruley, Robert 35	Millett
Breman, J. H. 31	"
Blakesley, H. 6	Grand Ledge
Bergin, William 31	Lansing
Bank, Joseph 24	"
Barrier, David 4	Grand Ledge
Benjamin, Darwin 10	Delta
Briggs, P. B. 7	Grand Ledge
Berger, John 7	"
Barr, Roswell 2	Delta
Bertram, Frank 25	Millett
Bertram, Mrs. Joseph 35	"
Barnes, Geo. 33	"
Bush, A. D. 13	Lansing
Casper, V. E. 3	Delta
Casper, Ida 2	"
Cummings, A. J. 3	"
Caldwell, John 21	Grand Ledge
Campbell, Henry 8	Delta
Crosby, A. C. 8	"
Campbell, Clarence 1	"
Critchet, L. H. 1	"
Cowen, James 20	Grand Ledge
Crouch, Frank 2	Delta
Cupit, Mrs. L. N. 1	"
Craw, Andrew 25	Lansing
Cowles, Henry 33	"
Clements, Henry 23	Millett
Clements, Geo. 23	"
Crouch, Caleb 11	Lansing
Crane, W. H. 35	Millett
Crane, Jonathan 35	"
Cole, William 35	"
Creyts, Miles 35	"
Carpenter, C. C. 26	"
Crumb, Ezra 10	Grand Ledge
Crumb, Mary 10	"
Cupit, Jno. 6	"
Cory, J. N. 14	Lansing
Darling, Geo. W. 20	Grand Ledge
Dunn, William 35	Millett
Dunlap, J. R. 36	"
Decker, Geo. 21	Charlotte
Dunn, William 35	Millett
Dunn, J. W. 25	"
Drake, E. L. 21	"
Digby, Chas. 24	"
Dayton, Samuel 14	Lansing
Easterly, Asa 21	Grand Ledge
Eldred, Wesley 32	Millett
Elliott, H. 17	"
Felton, A. Est. 1	Lansing
Flitton, John 34	Grand Ledge
Foster, Mary A. 13	Lansing
Felske, August 11	"
Fox, A. C., Sr. 8	Delta
Grinnell, Amelia C. 5	"
Fuller, G. O. 27	Lansing
Fishell, John 5	Grand Ledge
Felske, Chas. 9	"
Frost, Jason 9	"
Farran, John 6	"
Foster, William 14	Lansing
Foster, William A. 14	"
Fowler, Frank 13	"
Franks, James 17	"
Felske, Otto 11	"
Foster, Mrs. James 11	"
Fry, L. G. 14	"
Farnsworth, H. B. 5	Grand Ledge
Farnsworth, C. R. 5	"
Farnsworth, Mary A. 5	"
Fishell, Merritt 5	"
Fox, A. C., Jr. 36	Dimondale
Flanders, Martha 35	Millett
Ferrel, William 29	Dimondale
Garrett, M. K. 20	Lansing
Gillett, Myra 1	"
Gasson, G. B. 24	"
Grinnell, P. C. 17	Grand Ledge
Graham, Geo. 12	Lansing
Griffith Thomas 15	"
Goodrich, M. E. 2	Delta
Grattan, Harvey 3	"
Henderson, Joseph 10	Lansing
Hazel, Frank 16	Grand Ledge
Harpster, Mrs. James K. 20	"
Hamilton, John 33	Millett
Huxtable W. Lydia 11	Delta
Holloway, John 15	"
Hallock, E. E. 17	Grand Ledge
Hayden, Jasper 15	"
Huxtable, William J. 11	Delta
Huxtable Thomas 10 & 2	"
Holly, Chas. A. 22	Lansing
Hull, Mrs. M. E. 24	"
Hart, Edwin 8	Delta
Hatchell, Geo. 21	Lansing
Hadley, J. A. 1 & 2	Delta
Halstead, Andrew 20	Grand Ledge
Harpster, Ellen 3	"
Huddleston, H. 10	"
Hunter, Mrs. J. F. 26	Millett
Hazel, Eugene 17	Grand Ledge
Hume, A. S. 15	Lansing
Hazel, Frank 4	Grand Ledge
Hamilton, James 29	Millett
Hamilton, Zenas 18	Grand Ledge
Holbert, Amos T. 6	"
Heckman, Ellen 25	Lansing
Huxtable, John 13	"
Harpster, Daniel 5	Grand Ledge
Hutton, F. W. 1	Lansing
Hart, Geo. D. 23	Grand Ledge
House, Lucy 33	Dimondale
Ingersoll, Jane 20	Grand Ledge
Ingersoll, Elliott 8	Delta
Ingersoll, Harlow 8	"
Kirtz, Benj. 20	Grand Ledge
Kettle, Geo. 2	Delta
Ketcham Lucy 8	"
Kraft, Albert 10	"
Krieger, Henry 14	Lansing
Knapp, Mary L. 33	Millett
Knapp, C. E. 6	Grand Ledge
Kennelly, John 31	"
Knight, Albert 31	Lansing
Lemon, Chas. 35	Millett
Lemon, Mrs. Geo.	"
Lamphere, Lyman 35	"
Liddle, A. H. 35	"
Lawrence, S. D. 18	Grand Ledge
Lowell, Lillian 8	Delta
Lemon, Elton 31	Millett
Lee, John H. 4	Delta
Lamerson, W. 27	Lansing
Lee, Geo. N. 9	Delta
Lee, John 26	Millett
Leavitt, Waterman 10	Delta
Lade, W. 21	Lansing
Loomis, James 8	Delta
Mascho, Henry 18	Grand Ledge
McGiee, J. B. 21	Millett
Munton, Isaac 35	Grand Ledge
McDonald, Fred B. 1	Lansing
Miller, S. S. 6	Grand Ledge
Minor, Wooleot 3	Delta
Mosher, C. R. 11	"
McCaple, Frank 31	Grand Ledge
Maier, Christian 6	"
Maier, Martin Mrs. 6	"
Martin Eliza 8	Delta
Mosier, William 2	"
Munson, J. B. 34	Millett
Mason, A. W. 23	Lansing
McKelvey, A. 31	Dimondale
McClaplin, John 20	Grand Ledge
McKenzie, Thomas 34	Millett
Maier, Geo. 5	Grand Ledge
Mann, F. 16	Lansing
McRoberts, William 3	"
Moore, Cyrus T. 30	Grand Ledge
Maier, C. F. 4	"
Monroe, Orlow 10	Delta
Moon, Mrs. M. W. 29	Lansing
Murray, John, Estate of 12	"
Moon, H. H. 34	Millett
Myers, S. P. 10	Lansing
Nixon, Samuel 17	Grand Ledge
Nolber, Henry 1	Lansing
Openlander, William 9	Delta
Openlander, Geo. 15	"
O'Connor, Dennis 24	"
Parker, Elsworth 20	Grand Ledge
Parker, Earnest 29	"
Purcell Andrew 25	Lansing
Parfree, G. D. 19	Grand Ledge
Perkins, John 29	Millett
Perkins, Geo. 35	"
Plowman, Luther L. 8	Ingersol
Perkins, Henry 33	Millett
Potter, Solomon 8	Delta
Parker, D. M. 14	Grand Ledge
Phillips, Geo. 3	Delta
Parsons, A. T. 21	Lansing
Perkins, Charlotte 26	Millett
Parker, Cornelius 8	"
Parker, Norman 17	Grand Ledge
Parker, A. L. 5	Delta
Parker, J. H. 29	Grand Ledge
Patterson, Mary 20	Lansing
Richmond, William 35	Millett
Ruhl, Mrs. Lawrence 21	Grand Ledge
Ruhl, Joseph 19	"
Reed, Susan D. 35	"
Reeve, Alfred 12	Lansing
Ruesser, P. B. 10	Delta
Rogers, M. 31	Millett
Ring, Ira 6	Grand Ledge
Rander, P. F. 6	"
Roundville, Chas. 2	Delta
Rouse, John 35	Millett
Robbins, Henry 12	Lansing
Robinson, J. R. 19	Grand Ledge
Reeve, William 4	Delta
Roe, William 28	Grand Ledge
Soper, James 3	Delta
Sanders, E. P. 3	"
Spangler, Mrs. A. 3	"
Space, James B. 4	"
Smith, Mrs. J. H. 5	"
Sammis, James 31	Millett
Searman, H. H. 35	"
Sherman, W. D. 11	Delta
Stafford, Alonzo 8	"
Stirling, Delia 6	"
Smith, Hannah 8	"
Sheets, Geo. 22	Grand Ledge
Shepler, John 17	"
Slaughter, L. 22	Lansing
Slaughter, A. D. 27	"
Sherwood, Henry 21	"
Smith, Amos 14	"
Simons, James 18	"
Simons, C. A. 23	"
Stowell, Wm. 22	"
Seiler, William 31	Millett
Stafford, P. A. 3	Delta
Shipman, Mrs. E. D. 7	Grand Ledge
Smith, W. H. 8	"
Schneeberger, Fred 14	Lansing
Selby, Herman 34	Millett
Soper, Joshua 9	Delta
Space, John N.	Grand Ledge
Sutliff, L. T. 18	Lansing

DELTA TOWNSHIP.

Sutliff, Geo. 12	Lansing
Snow, J. H. 28	Millett's
Sanborn, E. P. 17	Grand Ledge
Shipman, S. W. 7	"
Shipman, Mrs. C. M. 7	"
Smith, Ira J. 7	"
Stabler, Chris. 36	Lansing
Seier, John 36	"
Smith, E. P. 30	"
Smith, R. E. 6	Grand Ledge
Smith, B. B. 5	"
Shaller, F. W. 16	Delta
Soper, Mason 10	"
Smith, B. F. 15	"
Smith Hem. 6	Grand Ledge
Smith, Simeon 5	"
Spare, W. J. 9	Delta
Sheets, Irving 17	Grand Ledge
Stafford, Geo. 9	Delta
Sawyer, Geo. W. 4	"
Swan, H. W. 48	Millett's
Shappell, Jacob 33	"
Shank, E. H. 16	Grand Ledge
Shindlinger, C. L.	Lansing
Sessions, Jane L.	"
Schultz, Fred 29	Grand Ledge
Sharp, E. C. 11	Lansing
Schler, Fred 26	"
Strong, Geo. E. 22	"
Spare, N. H. 30	Grand Ledge
Tower, Alex 15	Lansing
Tower, J. A. 15	Delta
Taylor, Chas. 9	Grand Ledge
Taylor, David 4	"
Thomas, Elizabeth 35	Millett's
Taylor, Jane 5	Grand Ledge
Tower, J. B. 6	"
Underhill, Jane 35	Millett's
Underhill, Levi 35	"
Underhill, Henry 35	"
Vanderwalker, Mrs. A. E. 8	Delta
Vanderwalker, Ezra 8	"
Winters, John 14	Lansing
Wright, Joel 5	Grand Ledge
Wilson, Frank 3	Delta
Walters, Andrew 35	Millett's
Williams, Eugene 1	Lansing
Wardell, J. A. 34	Millett's
Wells, Mary 19	Grand Ledge
Welch, William 31	"
Waldo, A. N. 15	Lansing
Waldo, Mrs. H. A. 6	Grand Ledge
Wells, E. J. 19	"
Walker, Jonathan 36	Lansing
Woodard, Gardner 3	Delta
Wall, G. H. 35	Millett's
Wedge, William 25	Lansing
Ward, Thomas	Millett's
Waldron, John 35	Millett's
Wall, John 25	Lansing
Warren, Frank 4	Grand Ledge
Woodward, Marvin 35	Millett's
Wardell, E. C. 35	"
Wardell, Frank 36	"
Watson, Fuller 8	Delta
Yerkes, Wm. E. 7	Grand Ledge
Yerkes, Wm. 8	Delta
Zachariah, Geo. 29	Millett's

EATON TOWNSHIP.

Allen, John F. 8	Charlotte
Allen, Henry, 8	"
Allen, Andrew 16	"
Allen, Homer 16	Eaton Rapids
Allen, Mrs. Harry 10	Charlotte
Allen, Horace 10	"
Alspaugh, John G. 31	"
Anglin, A. L. 31	"
Aulls, Chas. 12	"
Austin, S. M. 13	"
Barr, Geo. 1	Potterville
Bailey, Fred 11	Charlotte
Ball, Jesse 20	"
Ballard, Chauncey 34	"
Bellows, H. P. 25	"
Bennett, Charlotte 36	"
Bennett, A. L. 25	"
Bennett, Hannah 25	"
Blatt, Anthony 15	"
Blatt, Sarah 18	"
Blatt, Oiled 18	"
Blowers, James 14	"
Blackwell, Sarah 36	"
Bishop, Frederick 6	"
Boyer, S. S. 27	"
Brymer, R. F. 27	"
Britten, Cordelia 14	"
Britten, Chauncy 14	"
Britten, A. M. 15	"
Casler, John 24	"
Casler, Mary F. 24	"
Caufield, Elizabeth 15	"
Curtis, Horace 1	"
Cook, Eli 21	"
Cook, Elizabeth 34	"
Cobb, Henry E. 35	"
Cooley, Geo. 29	"
Cooley, Benj. 29	"
Casler, Henry 10	"
Casler, Hiram 9	"
Crane, R. F. 21	"
Crane, Amasiah 24	"
Clay, James H. 29	Charlotte
Cooley, W. T. 32	"
Cooley, W. H. 28	"
Cox, Lucinda 23	"
Cox, Merritt 23	"
Crosier, Irvin 35	"
Crosier, Chas. 23	"
Clay, D. H. 33	"
Curry, Dan 15	"
Curry, Latimore 14	"
Childs, James 1	Potterville
Davis, Mrs. John W. 25	Charlotte
Depue, James M. 14	"
Dorman, John 26	"
Dorman, Lemuel 35	"
Dickinson, Daniel 21	"
Dickinson, L. D. 21	"
Eddy, Dennis 11	"
Fay, Jane A. 14	"
Fields, Chas. 25	"
Fields, Albert 35	"
French, James 34	"
Foster, Frank 30	"
Flynn, William 10	"
Fast, Eliza 1	Potterville
Fast, Jonas 1	"
Fast, Clinton 1	"
Freeman, Chauncey 10	Charlotte
Foster, Wesley 5	"
Garvey, James 2	"
Garvey, William 2	"
Garvey, Thomas 11	"
Gidner, Margaret 3	"
Gidner, Agist 3	"
Greenman, M. A. 11	"
Gerak, J. B. 9	"
Griffin, James E. 21	"
Gibbs, H. A. 36	"
Gibbs, Chester 21	"
Gibbs, Ellen 21	"
Gibbs, Martin 11	"
Hill, Mary A. 1	Potterville
Hartell, Lewis 2	"
Hartel, Frank 2	"
Hauffner, Andrew 31	Charlotte
Hauffner, Chas. 31	"
Hipp, G. F. 22	"
Huber, Ezra 25	"
Hart, Thomas 12	"
Howard, Stephen 11	"
Heffner, Reuben 15	"
Hale, Omar 32	"
Holcomb, Harvey 23	"
Hazard, Gouldie 27	"
Hoey, A. W. 31	"
Hoey, Irvin 31	"
Holcomb, LeRoy 21	"
Holcomb, Harry 21	"
Holcomb, Annet 21	"
Hall, Merrick 16	"
Hale, Kara B. 28	"
Jewett, Chauncey 26	"
Kirk, Albert 31	"
Kint, Aaron 33	"
Knapp, Geo. B. 16	"
Kinsell, G. W. 5 & 8	"
King, J. B. 5	"
Leibold, C. H. 25	"
Long, Konrad 5	"
Leacher, John 24	"
Long, John 26	"
Laverty, Cyrus 1	"
Linsley, Abigail 1	Potterville
Long, Elizabeth 14	Charlotte
Long, William 14	"
Linsley, Will 21	"
Lamb, Eli 21	"
Lloy, Anne 16	"
Laverty, Hiram 9	"
Laverty, G. W. 4	"
Miller, Josiah 23	"
Miller, Cornelius 19	"
Milbourn, Frank 1	Potterville
McConnell, T. F. 6	Charlotte
Morse, O. D. 2	"
Marshall, Alonzo 4	"
Murray, James 29	"
Miller, Willis J. 33	"
McFarland, Ann 27	"
Mervel Michael 23	"
McClintic, Joseph 30	"
McClintic, G. W. 10	"
Miller, William 23	"
Milbourn, Amanda 13	"
Moore, Joseph 26	"
Milbourn Fred 11	"
Newth, Thomas 11	"
Newth, Jacob 11	"
Perkey, Henry 28	"
Puffenberger, Sarah 27	"
Pierce, William 14	"
Puffenberger, Lyda 27	"
Pixley, James 28	"
Perkey, Hannah 35	"
Parker, Mrs. Agusta 6	"
Parker, Arthur 6	"
Parker, Fred 6	"
Pierson, Miles 34	"
Perkey, Alfred 4	"
Quantrell, William, Sr. 19	"
Quantrell, William, Jr. 19	"
Richards, Allen S. 1	Potterville
Richardson, Frank 9	Charlotte
Rice, M. S. 15	Charlotte
Rochester, Frank 21	"
Randall, Hattie 29	"
Royston, Cornelius 13	"
Riley, H. O. 21	"
Richardson, Frank 9	"
See, Abram 31	"
See, Eldridge 29	"
See, James 27	"
Sheaver, Harriett 31	"
Stoneaker, Ed. W. 27	"
Stewart, Julius 5	"
Sabin, Elmer 4	"
Summers, Adam 4	"
Stewart, Benj. 4	"
Sherman, Alexander 12	"
Southworth, Alonzo 4	"
Southworth, Chas. 4	"
Southworth, Susan 34	"
Southworth, Allen 17	"
Southworth, M. 20	"
Stine, W. W. 21	"
Searles, Sarah 19	"
Searles, John 17	"
Shell, Isaac 27	"
Spotts, Benj. 28	"
Snow, Thomas J. 19	"
Smith, Allen A. 23	"
Smith, W. S. 25	"
Snoke, Abram 35	"
Smith, E. M. 23	"
Smith, Jacob 21	"
Solomon, Conrad 26	"
Tubbs, Lafayette 30	"
Thompson, Geo. 15	"
Upright, D. C. 21	"
VanDeusen, James 33	"
VanDeusen, John 34	"
VanDeusen, Harvey 23	"
Wells, Jefferson 14	"
Wack, Elizabeth 11	"
Woodman, Henry 18	"
Wells, David 9	"
Wickard, Daniel 22	"
Wickard, Mervin 20	"
Ward, Mrs. Geo. 18	"
Wright, Elson 16	"
Wright, Nathan 24	"

EATON RAPIDS TOWNSHIP.

Abbey, C. 12	Eaton Rapids
Abbey, Y. J. 11	"
Adams, Milton 13	"
Alderman, Wm. 10	"
Albina, Lewis 8	Kingsland
Annis, Morris 28	Eaton Rapids
Annis, S. A. 29	"
Annis, James 18	"
Andrews, J. L. 27	"
Arnold, Nancy 21	"
Arnold, H. B. 21	"
Arnold, J. H. 21 & 18	"
Ashley, Mary N. 15 & 26	"
Ashley, Isabella 3	"
Ashley, Samuel 8	"
Bancroft, J. D. 21	"
Barrett, C. L. 14	"
Barrett, F. S. 13	"
Barrett, Ora 35	"
Bartlett, C. M. 8-19	"
Bates, E. 1	"
Beilly, Geo. 21	"
Beadle, Rachel 2	"
Bement, G. M.	"
Bement, L. 12	"
Bement, Wm. L. 2	"
Bennett C. E. 5-6	"
Bellows, Warren 1 & 2	Kingsland
Bentley, Benj. L. 19-33	Eaton Rapids
Bentley, John W. 21-28	"
Bentley, Oscar E. 20	"
Benton, Wm.	"
Blodgett, Rich	"
Boody, Mrs. A. J. 7-8	"
Boody, Wm. 7-8-9	"
Bodell, Jas. 6	"
Booth, Geo. 33	"
Boteler, Chas.	"
Boteler, James 31	"
Boteler, Fred E. 31	"
Bradford, A. L. 27	"
Brandmore, E. 5	Kingsland
Brainard, Frank M. 27	Eaton Rapids
Bray, H. A. 3	Kingsland
Bramble, Harrison 14	Eaton Rapids
Brewer, Angeline 21	"
Britten, C.	"
Brightman, Ellen N. 19	"
Brown, Andrew	"
Broughton, N.	"
Brunton, Robert 10	"
Bromiley, P. K. 11	"
Balcock, Emery L. 5	"
Balcock, H. 5	"
Bonker, Geo. W. 24	"
Bauber, Chas. 18	"
Bush, A.	"
Bush, Willis 36	"
Cady, Mrs. Geo. 15	"
Canada, Jno., Est. 12-14	"
Canada, A. 14	"
Canfield, F. A. 9-12-16-6-16	"
Carr, C. L. 24-26	"
Casler, M. J. 27	Eaton Rapids
Casler, Benj. 16	"
Cidercolter, Mrs. M. 21	"
Clark, Geo. 16	"
Clark, James 25	"
Conklin, Lee 10	"
Corey, James 20	"
Crandall, J. H. 3	"
Crocker, H.	"
Crocker, L. A. 11	"
Cross, H. 21	"
Cummings, Jesse 1	"
Cupp, W. C. 17-18-19-20-30-29	"
Curtis, M. J. 35	"
Daly, C. W. 10 & 33	"
Darling, A. C. 26	"
Darling, Simon 2	"
Decker, Elizabeth 25-26-35	"
Densmore, Jno. P. 15	"
Depue, Daniel 4	"
Depue, Geo. 4	"
Depue, Milton 6	"
Depue, Wilson 9	"
Dillingham, Mrs. Stephen 11	"
Dillingham, A. 10	"
Dobson, Rosella 2	"
Dodge, Jno. S. 35	"
Dowmaner, A. G. 10	"
Dwinnell, S. T. 29	"
Edmonds, L. M. 29	"
Ferris, G. W. 1	"
Ferris, Job 26	"
Ferris, Wesley 26	"
Freer, James 19	"
Freer, R. A. 21	"
Flaherty, E. 15	"
Ford, Alice 19	"
Gallery, Geo. 35	"
Garvey, Cath. 7	"
Gillette, A. R. 8-7	"
Gillette, R. 10	"
Goldman, Jno. 27	"
Green, Alvin	"
Green, W. P. 23	"
Griffin, G. C. 2	"
Griffin, G. P., Est. 11	"
Griffin, Horace 22	"
Griffin, Lorenzo 21	"
Griesenberger, T. 20	"
Griswold, Chas. H. 20	"
Groesbeck, James 16	"
Groesbeck, Daniel	"
Groesbeck, Geo. 16	"
Gunnell, Wm. 9	"
Halifax, Ed 9	"
Halifax, Fred 10	"
Hoyt, Mrs. E. 11	"
Hartenberg, Chas. 2	"
Hicks, Geo. W. 1	"
Hicks, Liberty 27	"
Holcomb, Sally 11 & 15	"
Holcomb, Horace 26	"
Hendershott, A. 15	"
Hunt, C. M. 21	"
Haskins, Lucy A. 28	"
Harrington, Elizabeth	"
Haywood, W. A.	"
Hodges, Luther 21	"
Herrick, O. 26	"
Hall, F. P. 22	"
Holbrook, Mrs. M. A. 8	"
Hoyt, H. P. 26	"
Hamlin, D. H. 25	"
Johnson, Miles 19	"
Jack, W. A. 26	"
Kelley, T. P. 1 & 12	"
Kelley, Wm. 11	"
Kenney, M. F. 21	"
Kint, W. H. 31	"
Kifer, John 15	"
Klotz, David 20 & 21	"
Lavcock, Wm. 24	"
Lamb, Ross W. 20	"
Lamb, F. D. 33	"
Lapham, H. C. 24	"
Leach, Anna M. 22	"
Leisenring, Jno. W. 21	"
Lewis, Sam 13	"
Long, Mrs. F. T. 31	"
Long, S. A. 25	"
Long, Ellsworth 21	"
Lockwood, Jas. 8	"
Mahoney, M. 22	"
Manning, J. 29	"
Martin, Chas. 1	Charlotte
Martin, H. F. 1	"
McManus, Elizabeth 28	Eaton Rapids
McCullough, Thos. 21	"
Milburn, Peter 10	"
Milburn, Fred 7	"
Miller, Robt. 19	"
Mix, S. C. 20	"
Moon, Harriet 8	"
Mordoff, M. 17	"
Moyer, John 3	"
Montgomery, Mrs. J. S. 23	"
Moore, J. H.	"
Myers, Mrs. R. H. 35	"
Nadell, Clarence 17	"
Narlcon, Jas. 19	"
Nelson, Reuben 34	"
Norton, M. E. 1-3-14-4 & 6	"
Norton, Geo. W. 13	Eaton Rapids
Norton, J. R. 11	"
Norton, Walter 5	"
Olmstead, S. C. 11	"
Olney, R. 31	"
O'Leary, H. 22	"
Osburn, E. C. 24-25	"
Owens, Wm.	"
Owens, Thos. 24	"
Overshier, E. F. 26	"
Oxendale, Geo. 4	"
Parker, Mrs. A. 22	"
Parker, Guy C. 25	"
Parker, Frank 35	"
Peet, Chas. E. 10	"
Peck, Sophrona 21	"
Phelps, Lucy 5	"
Phillips, Wm. C. 31	"
Phillips, Chas. E.	"
Phillips, E.	"
Pfister, Grant 8	"
Pike, Wm. 23	"
Pickworth, Thos. 34	"
Polhemus, Mary 27 & 28	"
Porter, Albert 28	"
Potter, Mrs. D. H. 2	"
Powers, Calvin 34	"
Puffenbarger, Geo. 20	"
Pugsley, S. V. 21	"
Purdy, Mrs. S. 35	"
Ranney, Carey, 20	"
Raymor, Chas. 1	"
Royston, Wm. 16 9	"
Royston, Geo. H. 9	"
Rae, Wm. 31	"
Rogers, Levi 15	"
Rogers, Seth 15	"
Rogers, Prudence, 4	"
Rogers, Chas. 26	"
Rogers, Sarah 18	"
Rogers, N. H. 26	"
Root, S. 8	"
Sanders, Chas. 1	"
Saxton, A. D. 14	"
Schofield, J. 11	"
Schofield, Geo. H. 23	"
Shaul, H. C. 31	"
Sherman, H. H. 1	"
Sheldon, G. W. 9	"
Songtown, Jas. 36	"
Seelye, J. W. 27	"
Seelye, Steel 27	"
Smith, J. 29	"
Smith, Marshall 14 & 23	"
Smith, Eugene 15	"
Smith, C. M. 20	"
Smith, Albert 26 & 27	"
Smith, John W. 3	"
Smith, Levi 13	"
Spear, Mariah	"
Spear, David 2	"
Spears, E. H. 19	"
Spears, Perry 19 & 20	"
Spencer, Wm. 10	"
Spotts, Daniel 26	"
Spicer, Wm. 27 & 25	"
Sprague, H. C. 31	"
Springer, O. H. P. 26	"
Springer, Electa 35	"
Samoria, W. 1	"
Samoria, Wm. 1	"
Swigman, John 4 & 5	"
Stevenson, John 2	"
Stevens, C. W. 28-29-30	"
Sterling, Wm. 21	"
Sterling, F. R. 21	"
Sweet, Wm. A. 5 & 8	"
Terrill, D. H. 1	Kingsland
Tiffany, H. 2	Eaton Rapids
Thompson, Myron 22	"
Thompson, Mrs. Geo. 10	"
Travis, Jane 17	"
Trank, J. J. 5	Kingsland
Turner, Fred 12	Eaton Rapids
Vanfleet, James 15	"
Ward, Wm. 28	"
Warren, Chancellor 14	"
Welch, S. J. 14	"
Welch, Erwin 6	Charlotte
Wellman, W. S. 3	Eaton Rapids
White, Alonzo 18	"
Winslow, Frank 3	Kingsland
Winslow, Percy 8	"
Wilbur, W. S. 2	Eaton Rapids
Willmer, C. H. 18	"
Wilcox, Jas. D. 12-11	"
Wilkinson, Chas. 3	"
Williston Fred H. 33	"
Winchester, S. H. 20-21	"
Williams, C. E. 24	"
Wise, Lee 25	"
Wills, O. 26	"
Wood, Marshall, 21-21	"
Wood, Mrs. Julia 21 & 22	"
Wood, Thos. 23	"
Wood, Henry 21	"
Wood, J. M. 21 & 3	"
Ward, Randall 4	"
Young, Daniel J. 4-5	"
Young, Geo.	"
Young, Eel	"
Zentmyer, O. H.	"

KALAMO TOWNSHIP.

Ackley, Albert 8	Kalamo
Ackley, Warren 21	"
Armstrong, W. 29 & 30	"
Angell, Dianah 21 & 28	"
Ackley, Andrew 18	"
Andrews, John 6	Nashville
Andrews, John 7	"
Benedict, Stephen 5	"
Barnes, T. C. 5	"
Barnes, Porter 5	"
Brown, Duane 6	"
Brown, John A. 6	"
Brown, Eugene 18	"
Buck, Phebe 19	Maple Grove
Bradley, T. M. 36	Alager
Benson, Isaac 4	Vermontville
Berland, Andrew 4	"
Briggs, Mrs. Abel 2	"
Briggs, Ed. 1	"
Briggs, Alonzo 1	"
Briggs, Carmel 3	"
Beaudiet, Chas. 3 & 4	"
Bordeaux E. 13	Carlisle
Brandell, Fred 13	"
Bromberg, Gustaf 17	"
Baker, Nathan 17	"
Bartig, Frank 3 & 12	"
Briggs, Nelson 14	"
Baker, C. H. 17	Kalamo
Baker, W. A. 17	"
Brundige, William 18	"
Bowen, D. G. 28, 29 & 32	"
Bowen, E. E. 28, 31, 32 & 33	"
Boyd, Chas. 15	"
Baker, E. M. 20	"
Babcock, Fred 29	"
Bracey, O. D. 33	"
Baker, John 34	"
Bissell, Lucy 22	"
Babcock, Frank 21	"
Bowen, George 22	"
Bruce, L. H. 27	"
Bowen, Frank 33	"
Bailey, Henry 27	"
Baker, B. F. 27	"
Clemens, Joel 22	"
Colton, Luther	"
Crouse, Lavina 31	"
Chambers, James 31	"
Conley, C. D.	"
Collins, C. G. 10	"
Courtright, J. W. 23	"
Campbell, J. F.	Carlisle
Curtis, Levi 20	Kalamo
Cronhurst, R. H. 10	"
Collins, Elizabeth 16	"
Cole, E. H. 24	"
Cole, Frank 21	"
Cunningham, William 1	Chester
Cosgrove, Fred 22	Ceylon
Conklin, C. H. 43	"
Courtright, John 18	Carlisle
Cooper, C. A. 24	"
Conklin, J. H. 25	"
Coubble, Mary 26	"
Campbell, G. A. 26	"
Curlew, Clara 21	"
Curlew, Frank 24	"
Dutes Clara 35	"
Dean, Silas 12	"
Day, N. E. 36	Alager
Dibble, L. H. 31	Ceylon
Dye, Etta 31	"
Davis, William 7	Nashville
Donaldson, Mrs. G. 17	"
Donaldson, Gilbert 8	"
Dean, Martha 19	Kalamo
Driscoll, D. W. 28	"
Dotson, Joseph 29	"
Donaldson, Mary 18	"
Daugherty, A. D. 22	"
Dodgson, John 13 & 15	"
Ellis, Sarah 10 & 16	"
Eason, John 17	"
Earl, J. M. 28	"
Ehret, John 20	"
Ellis, Bernard 10	"
Eaton, Angeline 13	"
Ellis, F. W. 14 & 23	"
Ellis, E. K. 18	"
Edmonds, Samuel 23	"
Ehret, Nora 8	Nashville
Finch, Gertie A. 6	"
Fitch, Dell 8	"
Fowler, Mrs. J. 17	Kalamo
Fowler, William 17	"
Fowler, Mrs. J. 21	"
Fowler, J. M. 21 & 31	"
Flatham, Bertha 4	"
Flatham, Carl 4	"
Fowler, J. M. 16	"
Fitch, Dell 8	Nashville
Fails, G. O. 36	Carlisle
Frighner, Frank 28	"
Fisher, William 18	"
Gage, Lafayette 25	Kalamo
Gould, Mrs. J. J.	Kalamo
Gridley, Reuben 21, 29 & 31	"
Gridley, E. H. 21	"
Gridley, Mrs. Hiram 21 & 28	"
Graves, Carrie 28	"
Grant, Mrs. M. M. 12	"
Grant, Walter 22	Kalamo
Gregg, T. N. 22	"
Gridley, Orris 32	"
Greeman, Fred 25	"
Granger, W. F. 21 & 28	"
Gould, J. J. 22 & 24	"
Gould, J. V. 23-24-26.	"
Gould, O. G. 31 & 27	"
Green, William 17	"
Griffin, Adaline 6	Nashville
Griffin, Frank 6	"
Granger, Mary 1	Vermontville
Gehman, Cyrus 3	"
Gusey, C. 1	Chester
Gusey H. 1	"
Hall, Emily 9 & 22	Kalamo
Hoffman, John 15	"
Harr, Mrs. Eben 21	"
Harmon, Lena 23	"
Hall, T. O. 22	"
Hayden, Geo. 27	"
Herring, Chas. 21	"
Holman, C. L. 31	"
Hall, Edward 8 & 10	"
Hand, John 16	"
Hoover, Parmelia 8	"
Harmon, Hannah 1	"
Hunter, William 4	Vermontville
Hills, Geo. 3	"
Horton, Henry 4	"
Hartwell, Frank 17	Nashville
Hire, Royal 13	Carlisle
Hall, S. O. 21	"
Hubbard, L. H. 4	Vermontville
Harwood, W. H. 24	Carlisle
Imes, Andrew 32	Bellevue
Imes, Simeon 32	"
Hill, E. W. 1	Chester
Johnson, W. F. 34 & 35	Kalamo
Johnson, Elizabeth 35	"
Johnson, Berwick 21	"
Jacobson, John 9	"
Johnson, Henry 5	Nashville
Kelsey, Minerva 19	Kalamo
Kent, Leander 21 & 22	"
Kelsey, Ira A. 30	"
Lyon, T. J. 23 & 26	"
Lyon, W. L. 23	"
Lydy, Samuel 31	"
Lindsley, Ira S. 16	Carlisle
Link, Andrew 11	"
Leonard, J. H. 28	"
Morris, M. G. 7-8	Kalamo
Mason, William 7-8-16	"
Mattison, E. G. 29	"
Mead, Willard 33	"
Mead, O. E. 35	"
Mead, O. W.	"
Mead, Daniel 34	"
Mast, Benj. 19	"
Morris, Chas. 28	"
McGinnis, Harvey 29	"
Mason, A. H. 22	"
Merriam, S. S. 30 & 29	"
Means, R. P. 29 & 33	"
Mapes, Ira 31	"
Mattison, S. O. 36	Ceylon
Martins, John 32	Bellevue
Martins, Geo. 33 & 35	"
Murray, John 30	"
Martins, William 30	"
McCarty, John 3-10 & 11	Vermontville
Matthes, Louis 10	Carlisle
Miller, Gustaf 1	"
Montgomery, Robt. 24	"
McKinnis, L. 6 & 7	Nashville
Nix, J. D. 6	"
Merritt, C. C. 7	"
Nix, Alberman 7-8 & 16	"
Nix, J. B. 7	"
Nix, John D. 7	"
Nix, Dean 7	"
Mattison, Asa 16	Nashville
Myers, Daniel, 10	"
Mason, Edward 16	"
McConnell, Sophia 36	Carlisle
Mead, Willard 35	Kalamo
McLaughlin, Margaret 13	"
Murch, Mrs. C. D. 24	"
Nye, P. H. 14	"
Nye, Mrs. P. H. 14	"
Nash, Edward 8	Nashville
Nelson, A. K. 22	Kalamo
Nelson, William 22 & 27	"
Ovenshire, Earl 31	Bellevue
Ovenshire, Bert 31	Kalamo
Oaster, W. H. 32	Bellevue
Ovenshire, Lafayette 33	Carlisle
Odell, Mrs. J. L. 33	"
Pinnock, Frank 11	"
Pinnock, Joseph 21	"
Porter, D. W. 24	"
Porter, Chas. 23	"
Powers, S. W. 21	Kalamo
Pease, William 29	"
Perry, S. A. 27	"
Perry, Chas. 27	"
Perkins, E. J. 26	"
Powers, Eugene 11 & 14	Carlisle
Pond, E. H. 6	Nashville
Parker, J. R. 5	"
Pope, A. C. 2	Vermontville
Rowell, A. J. 21 & 21	Kalamo
Rouse, Erwin 27	Kalamo
Rouse, Mrs. E 22	"
Reed, Henry 22	"
Rickel, John 20 & 21	"
Roberts, John 18	"
Reynolds, Jacob 19	"
Roberts, W. A. 20	"
Ripley, A. G. 20	"
Reid, Harriett 13	Carlisle
Rager, Adelia 14	"
Ramsey, Harriett 13	"
Rolf, T. J. 13	"
Roach, J. W. 5	Nashville
Sanders, H. C. 26	Kalamo
Sanders, Wm. 29	"
Snake, Daniel 22	"
Shuter, Chas. W. 23	"
Showalter, Frances 7	Nashville
Slosson, Cyrus 5	Kalamo
Shepherd, Silas 16	"
Shepherd S. A. 16	Nashville
Stocking, H. J. 18	"
Swift, E. C. 5 & 19	"
Smith, Geo. 8	"
Sheldon, Edwin 20	Kalamo
Sweet, Jay 29	"
Spire, Lyman 30	Ceylon
Spire, William 30	"
Streeter, Ruth S. 19	Kalamo
Scott, H. A. 29	"
Sprague, Lewis 10	"
Stout, W. P. 11	Vermontville
Sears, Millard 12	Carlisle
Stockwell, Ezra 8	Vermontville
Smith, Henry 14	Kalamo
Sprague, Mrs. Norman 11	Carlisle
Sprague, Bert 11	Kalamo
Sanders, B. A. 4	"
Sanders, S. C. 25	"
Sweet, Levi 27	"
Swift, A. B. 26	"
Swift, A. P. 25 & 35	"
Scott, O. J. 35	"
Shepherd, W. D. 15	"
Silverthorn, Mary 13	"
Simons, R. J. 22	"
Simons, L. Z. 15	"
Simons, H. D. 23	"
Shuter, Herbert 15	"
Shuler, Chas. 18	"
Sprague, Geo. 9 & 10	"
Spradlove, John 16 & 17	"
Smith, Daniel 25	Carlisle
Sauer, J. S. 36	"
Swift, James 25	"
Smith, Daniel 26	"
Tarbell, S. C. 16	Kalamo
Tomlin, Joseph 17 & 20	"
Tinker, W. P. 22	"
Tyler, John 10 & 11	"
Tomlin, Oliver 21	"
Treiber, John 18	"
Treiber, Jacob 18	"
Trowbridge, S. D. 10	"
Townsend, Lovina 7 & 18	"
Tomlin, John 17	"
Tisch, R. E. 16	Nashville
Tyler, Edgar 3	Vermontville
Tyler, Fred 3	"
Tyler, Edwin 3	"
Tubbs, E. S. 1	"
Tubbs, C. W. 11	Kalamo
Tubbs, L. P. 11	"
VanDyke, Isaac 18	"
Vandyke, E. 23	"
Van Fleet, Roy 21	"
Viemaster, Willard 23	Ceylon
Wark, Leander 29	Kalamo
Wilson, W. W. 9	"
Willis, C. 9	"
Wilton, L. W. 16	"
Wilton, A. L. 9	"
Wilson, Margaret 24	"
Wilson, Mrs. J. 10	"
Wilson, James 10	"
Wood, Eber 15	"
Wilson, Mrs. Geo. 33	"
Wyble, Geo. 19	"
Wuirig, E. 24	"
Wyble, Martha 14	"
Wyble, James 11	"
Winters, Rebecca 24	"
Williams, F. O. 15	"
Wilcox, W. C. 19	"
Wilkinson, S. 8	"
Williamson, S. G. 35	"
Webber, Otis 15	"
Webber, J. H. 10 & 15	"
Wilson, T. L. 27	"
Wilson, J. A. 33	"
Wilson, C. W. 27	"
Wilson, Henry 27	"
Wilson, Chas. 22	"
Wildt, Fred	"
Wark, Albert 31	"
Wright, Wm. 8	Vermontville
Warden, William 1 & 2	"
Wells, C. K. 2	"
Wells, E. D. 2 & 3	"
Wilson, James 4	"
Wells, Mrs. Eli 5	Nashville
York, Mrs. John 27	Kalamo

ONEIDA TOWNSHIP.

Abbott, Mary 25	Grand Ledge
Allen, Chas. G. 21	"
Aden, Mrs. C. 14	"
Althouse, John H. 26	"
Althouse, George 23	"
Arnold, Geo. W. 3	"
Bailey, W. E. 11	"
Bailey, Borne 6	"
Babel, W. L. 31	"
Bass, M. A. 22	"
Baldwin, R. 2	"
Backus & Son, H. 23	"
Barton, Mrs. M. 7	"
Barton, Lewis 8	"
Baughman, E.	"
Beck, C. 12	"
Benjamin, F. W. 13	"
Beckman, J. C. 28	"
Beaumont, Ed 21	"
Bentley, C. A. 29	"
Bigelow, C. S. 16	"
Bigelow, E. C. 16	"
Blake, P. B. 5	"
Bobier, Alonzo 30	"
Bowen, Chas. 7	"
Boroughs, Samuel 19	"
Bruce, Geo. H. 19	"
Brandle, Henry 2	"
Brown, Geo. H. 1	"
Brunger, Fred 15	"
Brunger, Wm. 21	"
Brunger, Jas. 21	"
Brunger, Edgar 11	"
Brunger, John 8	"
Brunger, Chas. 27	"
Brunger, T. R. 27	"
Burns, Garrett 14	"
Case, O. S. 22	"
Chipman, A. H. 33	"
Clay, John 30	"
Clayton, John 7	"
Clark, J. M. 26	"
Clark, Dobert 9	"
Cole, Deville 8	"
Cole, Byron 7	"
Cole, Peter, Est. 14	"
Cole, N. H. 22	"
Crist, Elmer 20	"
Davies, M. A. 21	"
Dewitt, F. J. 5	"
Dewitt, Jonathan 5	"
Dewar, Wm. 14	"
Diebro, H. L. 16	"
Diebro, Melvin 16	"
Dingler, Jacob 24 & 13	"
Dubois, Jas. 15	"
Dudley, M. P. 22	"
Dunn, Jas. 36	"
Earl, Carl 29	"
Earl, H. P. 29	"
Earl, Estella 27	"
Eddy, Ellas 6	"
Edwards, Z. E. 27	"
Edwards, Jane 21	"
Edwards, Elmer	"
Elkins, Jay 25	"
Elkins, Edward 25	"
Ewing, J. W. 21	"
Eyestone, Mrs. F. E. 26	"
Fenn, Michael 9	"
Fenn, Wm. 16	"
Fenn, Chas. 18	"
Fenn, Geo. A. 16	"
Fenn, Chris. 22	"
Fenn, Fred 16 & 17	"
Flemming, William 27	"
Flemming, Alfred 17	"
Flemming, Alt. 17	"
Flemming, Albert 17	"
Figg, Sylvester 19	"
Foster, Ira 1	"
Foster, L. N. 1	"
Foster, Darwin 20	"
Fults, Washington 6 & 1	"
Fuller, J. E. 18	"
Fuller, E. H. 14	"
Fuller, C. V. 28	"
Gates, Geo. W. 24	"
Greer, Mrs. Geo. 31	"
Glenn, Cris, Jr. 35	"
Glenn, Cris, Sr. 35	"
Granger, Mrs. S. H. 35	"
Green, Wm. G. 18	"
Griffin, M. M. 26	"
Gross, Fred 1	"
Guinan, Luke 18	"
Guilford, I. N. 34	"
Guilford, Ransom 2	"
Hahn, Walter 23	"
Hayes, Mrs. E. M. 1 & 2	"
Hancock, John 22	"
Hatch, Edward 20	"
Hazleton, J. F. 18	"
Hazleton, Wm. 20	"
Henry, A. J. 30	"
Hill, N. W. 13	"
Hiegnalt, Martin 15	"
Hixon, E. J. 1	"
Hixon, Emily	"
Hixon, Levi	"
Hixon, Myron 17	"
Hixon, Myron 20	"
Hixon, Nathaniel 5	Grand Ledge
Hitchcock, Wm. 29	"
Howard, John 9	"
Holmes, Aaron 5	"
Holmes, Sylvia 26	"
Holcomb, Ira P. 36	"
Hulce, Benj. 14	"
Ives, D. L. 9	"
Jerome, Susan 1	"
Jeane, Silas 19	"
Jenne, O. S. 13	"
Johnson, James 16	"
Johnson, W. R. R. 6	"
Jones, J. V. 7	"
Jones, A. R. 7	"
Jones, R. L. 4	"
Jones, C. H. 4	"
James, Chloe 9	"
Joslin, Lucius	"
Joslin, L. M. 4 & 23	"
Judson, L. 31	"
Kebler, Chris 2	"
Kenyon, D. O. 29	"
Kelsey, Milo	"
Kelley, J. C. 30	"
Krapp, E. 1	"
Lankton, D. E. 30	"
Lapareel, Joe 34	"
Lawson, Mrs. E. 8	"
Lester John 22	"
Lewis, John 18	"
Lewis, Lucinda 34	"
Lewis, F. C. 10	"
Lewis, J. H. 10	"
Lamore, Frank 31	"
Major, Phillip 8	"
Mable, John 30	"
Marcho, N. S. 3	"
Maltby, C. S. 24	"
Mason, Horatio 31	Roxana
McMullen, E. J. 27	Grand Ledge
McMullen, Jennie 28	"
McMullen, J. M. 27	"
Merritt, Ed. 6	"
Miner, Glen 8	"
Miller, C. J.	"
Miles, Eli 16	"
Mitchell, Fred 15	"
Mitchell, Newman	"
Mitchell, Andrew 19	"
Mitchell, Chas. 19	"
Minnick, Wm. 23	"
Mitchell, Alfred 25	"
Mosher, Ed. 19	"
Moon, Marian 27	"
Moss, J. K. 3	"
Murray, John 26	"
Murray, A. J. 26	"
Newkirk, P. 12	"
Newton, Mrs. M. 13	"
Nixon, L. J. 35	"
Nixon, Ed 35	"
Nichols, Martin 36	"
Nichols, G. W. 35	"
Nichols, Truman 35	"
Nichols, J. P. 36	"
Nottingham, Mrs. 30	"
Nocker, Oliver 20	"
Nocker, Catharine 30	"
Oark, Wm. 17	"
O'Neil, J. H. 19	"
O'Neil, Allen A. 19	"
Otto, Jos. 23 & 34	"
Paine, Rowland 23	"
Patterson, C. A. 16	"
Pardee, Lorenzo 30	"
Parsons, S. L.	"
Parmenter, Philander 6	"
Parmenter, Fidelia 6	"
Parmenter, A. 9	"
Pelton, Wm. 17	"
Place, Daniel	"
Powell, J. W.	"
Preston, C. M. 23 & 26	"
Putterbill, Hattie	"
Quale, Ed. 23	"
Quantrell, A. B. 23	"
Reynolds, L. G. 10	"
Reynolds, A.	"
Reynolds, Rose 14	"
Reed, A. J. 14	"
Rice, Allen 13	"
Richardson, Chas. 30	"
Rogers, L. P.	"
Root, E. C.	"
Roth, Ed. 21	"
Rosier, Amos 29	"
Rosier, Russell 31	"
Russell, Wm. E. 12	"
Russell, Mrs. M. 24	"
Russell, J. W. 10	"
Russell, Fred 18	"
Russell, J. N. 10	"
Russell, A. C. 8	"
Salter, John 21	"
Salter, Martin 33	"
Sanders, Sam'l 5	"
Sanders, Geo. 5	"
Sanders, S. C. 5	"
Salisbury, E. J. 21	"
Schaal, Jacob 33	"
Sharp, Richard 24	"
Sauba, F. C. 23	"

ONEIDA TOWNSHIP.

Shultz, Henry 23	Grand Ledge
Sheets, M. R.	"
Sherman, Hart 31	"
Shepherd, L. H.	"
Skinner, F. Z. 29	"
Skinner, Rebecca 29	"
Singer, Geo.	"
Smith, Ira J. 23	"
Smith, Hiram	"
Smith & Smith 36	"
Smith, J. G. 23	"
Smith, Emiline 26	"
Smith, Angeline 21	"
Snyder, Geo. W. 13	"
Standish, Miles 23	"
Starks, John 19	"
Strange, J. S. 14	"
Strange, Daniel 34	"
Somerville, Robt. 11	"
Somerville, Frank	"
Sutherland, Peter 17	"
Sutherland, O. 17	"
Sutherland, Carson, 5 & 8	"
Sutherland, H. D. 8	"
Sutherland, Elmer 8	"
Sutherland, Ellha 8	"
Sutherland, Erick 8	"
Sutherland, W. H. 9	"
Sutherland, Lyman 4	"
Tanner, M. B. 22	"
Taylor, Eli 12	"
Taylor, T. V. 18	"
Thomas, M. V. 32	"
Tinkham, H. J. 25	"
Toms, Mrs. Thos. B. 5	"
Upright, Barbary 14	"
Van Alstine, Geo. W. 21	"
Van Alstine, L.	"
Vanloren, Daniel 14	"
Vogt, Fred 18	"
Voore, Ed. 20	"
Walsh, John H. 21 & 22	"
Wall, Chas.	"
Watson, W. J.	"
Watson, Geo. H. 20	"
Watson, N. G. 18	"
Watson, Elmer 17	"
Waldo, D. M. 13	"
Waldo, Silas 14	"
Waldo, C. J. 18	"
Waldron Est. 22	"
Whitney, Chas.	"
Wilkinson, F. M. 18	"
Williams, Ellsworth	"
Williams, James 20	"
Wiese, Emma L. 16	"
Wolpert, Homer 19	"
Wolpert, C. F. 19	"
Woods, Frank	"
Woods, Wallace 10	"
Worden, S. E. 8	"
Young, Geo. W. 8	"

ROXAND TOWNSHIP.

Allen, Albert 26	Hoytville
Allen, Geo. 26	"
Ablem, J. B. 16	"
Anderson, Mrs. Hans 29	"
Anderson, H. 29	"
Austin, Wallace 21	"
Austin, Hattie 8	"
Allen, J. 21	"
Aldro, A. 10	"
Anderson, Andrew 22	Roxana
Boyer, D. W. 22	Hoytville
Boyer, John 27	"
Boyer, Asa 22	"
Boyer, E. 20	"
Boyer, Peter 23	"
Boyer, W. 15	"
Boyer, O. M. 28	"
Brown, Zeroe 29	"
Brown, Catharine 29	"
Bemis, Albine 20	"
Bemis, Frank 21	"
Bennett, John 29	"
Bosworth, Leroy 20	"
Boyer, Jesse 26	"
Bishop, E. C. 3	"
Borton, B. 22	"
Bailey, Mrs. 10	"
Bailey, O. S. 20	"
Bosworth, William 30	"
Boyer, Morgan 15	"
Bennett, William 28	Bismark
Bird, Chas. 20	Roxana
Bosworth Bros. 31	Bismark
Beekman, H. F. 31	"
Bliss, Chas. 3	Mulliken
Brown, A. D. 3	"
Brown, H. M. 4	"
Brocout, Daniel 7	"
Barlow, J. E. 8	"
Barlow, Elias 10	"
Burroughs, Henry 5	"
Bailey, Roena 4	"
Boyer, Mary 3	Mulliken
Bolder, Milton 28	Roxana
Bobier, Joshua 21	"
Bedford, Henry 32	"
Cramer, Mrs. Emma 2	Mulliken
Cogswell, Alvin 2	"
Crane, Oscar 5	"
Conerite, Manley	Mulliken
Cryderman, Chas. 2	"
Cryderman, Frank 3	"
Crish, Elmer 35	Roxana
Crish, William 35	"
Carman, O. H. 26	"
Cogswell, Clark 35	"
Collins, James 16	"
Carman, Elizabeth	"
Caulkins, Melinda 17	Hoytville
Collins, Alvira 20	"
Cole, John S. 22	"
Caulkins, Mrs. H. 20	"
Codding, William 31	Bismark
Codding, Herbert 19	Hoytville
Dewey, M. H. 31	"
Dilley, Irwin 19	"
Dilley, Charlotte 19	"
Dilley, John 19	"
Dilley, Florence 19	"
Davis, Loyal 13	"
Davis, J. A. 23	"
Davis, V. E. 15	"
Davis, W. H. 21	"
Doyle, John 20	"
Depew, Amos 20	"
Dennis, C. E. 27	"
Dutcher, David 16	"
Dravenstratt, Mrs. 2	Mulliken
Davis, Walter H. 3	"
David, Chas. 3	"
Dunham, Chas. 6	"
Dickinson, H. S. 6	"
Doolittle, John 3	"
Dunham, J. W. 8	"
Dilley, Josiah 6	"
Doolittle, Chas. 10	"
Davis, Edwin 26	Roxana
Davis, Henry 26	"
Depew, Amos 26	"
Depew, Mary A. 24	"
Depew, David 21	"
Depew, Mrs. Sarah 21	"
Ewing, H. H. C. 30	Hoytville
Edgel, Henry 30	"
Edgel, Sylvester 22	"
Eastman, Hiram 1	Mulliken
Elsworth, A. E. 6	"
Eastman, Hiram	"
Eddy, Wallace 26	Roxana
Edmond, Mrs. Lucy 23	"
Fuller, Chas. 35	"
Ferris, Edward 22	"
Freeman, J. T. 22	Hoytville
Freeman, E. 22	"
Fitch, A. W. 1	Mulliken
Feightner, W. Mc. 3	"
Fisher, L. B. 5	"
Fonsie, W. B. 5	"
Gardner, Anna 1	"
Gardner, Mrs. A. 5	"
Gardner, Chas. 6	"
Gardner, S. E., Est. 9 & 10	"
Graves, Chas. 5	"
Gates, G. A. 5	"
Oaten, E. R. 4	"
Gardner, John 5	"
Grinnell, Mary E. 19	Hoytville
Gould, William 29	"
Green, Chas. 20	"
Green, C. F. 28	"
Granger, B. 18	"
Green, William 18	"
Green, Geo. 18	"
Gregg, B. H. 7	Sunfield
Granger, F. 31	Bismark
Hutchings, Martha 33	"
Haddix, Sloop 28	Roxana
Helm, D. V. 28	"
Hilliker, Dan'l 18	Grand Ledge
Hilliker, Geo. 12	"
Hamilton, J. N. 21	"
Hamilton, William 24	"
Hamilton, Chas. 26	"
Hiscock, Martin 13	"
Hosmer, E. 4 & 5	Mulliken
Hatch, H. B. 4	"
Hart, S. F. 6	"
Hill, J. J. 3	"
Hovey, John 8	"
Hale, John 5	"
Hay, Elmer 3	"
Halenbeck, David 5	"
Halenbeck, Mrs. Lucy 8	"
Hagerman, Benj. 9	"
Hart, Austin 7	"
Hagerman, Sam 5	"
Humphrey, John 21	"
Hart, H. F. 2	"
Haddix, Eliand 28	Hoytville
Houghton, Oscar 23	"
Hunt, A. E. 10 & 15	"
Holliday, Harlow 16	"
Hedge, S. C. 15	"
Howell, Wm. 35	"
Haddix, John 23	"
Hale, E. H. 23	"
Horton, Lemuel 23	"
Halladay, Mrs. S. 9	"
Halladay, Marlow 9	"
Harotte, C. W. 16	"
Haag, L. W. 27	"
Hollis, William 21	"
Ingram, William 8	Hoytville
Jones, Harris 3	Mulliken
Jones, Algood 3	"
Jones, Frank 8	"
Jackson, E. 35	Roxana
Jacobs, J. 28	"
Jackson, Mann 18	Hoytville
Jackson, Chas. 7	"
Jackson, Bert 10	"
Johnson, Samuel 5	"
Kesler, Edward 21	"
Kling, Henry 35	Roxana
Kingman, H. T. 28	"
Luther, Frank 28	"
Ludbrook, William 21	"
Ludbrook, W. W. 26	"
Litchfield, D. C. 19	"
Loucks, Jacob 20	"
Lamoka, Joram 20	"
Lamoin, Geo. 29	"
Litchfield, W. 22	"
Lundquist, Jacob 20	"
Lyon, Merritt D. 11	"
Larabee, Mrs. A. 7	"
Larabee, Martin 1	"
Lyon, James 17	"
Langtry, Samuel 8	Mulliken
Leigh, Mrs. H. 7	"
Ludbrook, Sanford 8	"
Lawrence, Albert 8	"
Lyon, Mrs. Sarah 8	"
Lockrow, Sarah 25	Roxana
McCumsey, Chas. 1	Mulliken
Marsh, W. E. 3	"
Merritt, Edward 8	"
Martin, A. T. 3	"
Morris, Floy 8	"
Mann, Elmer 9	"
Merritt, Frank 24	"
Morris, J. C. 8	"
Morey, Jno. 1	"
Merritt, E. A. 23	"
Mead, E. 24	Hoytville
Mead, Grant 14	"
Mead, Electa 14	"
Mead, C. 14	"
Mead, Irvin 14	"
Matice, L. B. 13	"
Mead, Harvey 14	"
Matice, Herbert 13	"
Markham, Herbert 27	"
Miller, Elisha 21	"
Moody, H. A. 18	"
Moody, S. 16	"
Moody, H. 16	"
Mead, James 13-14	"
Markham, Albert 27	"
Mead, Ezra 15	"
Moon, Spencer 25	Roxana
Moon, Francis 24	"
Meyer, Clem A. 24	"
McCargar, Chas. 24	"
McCargar, D. J. 35	"
Miller, Odra 36	"
Meyer, T. E. 22	"
Meyer, Leon 23	"
Miller, Geo. 30	Hoytville
Mead, William 21	"
Metzner, Jacob 7	Sunfield
Myers, Silas 31	Bismark
Merrifield, Geo. 5	Mulliken
Miller, Mrs. Mary 5	"
Newark, James 33	Hoytville
Nickle, Sabina 15	"
Newark, A. W. 22	"
Nickle, J. B. 10	"
Nickle, Sam 10	"
Nasher, Calvin 20	Roxana
Osman, Henry J	Mulliken
O'Neil, William 10	Hoytville
O'Neil, Henry 14	"
O'Neil, Mrs. C. 14	"
Osman, William 11	"
Ovit, Alexander 22	"
O'Neil, J. V. 15	"
Parker, Alfred 18	"
Parker, Jerry 17	"
Parker, John 17	"
Potter, M. 17	"
Potter, John 17	"
Pumphrey, Horton 20	"
Parker, Rufus 15	"
Parker, B. J. 22	"
Pitter, Edgar 17	"
Parker, John H. 16	"
Parker, A. L. 8	Mulliken
Peabody, S. 4 & 11	"
Peabody, Hiram 4 & 11	"
Plumb, Frank 8	"
Potter, Geo. E. 8	"
Peabody, Albert 5	"
Partlow, William 5	"
Pugh, Frank 21	Roxana
Preston, Ambrose 25	"
Quackenbush, Calvin 16	Hoytville
Root, Mrs. Elsie 25	Roxana
Rix, Geo. 35	"
Rix, Robert 25	"
Rix, E. B. 25	"
Rice, E. 34	"
Rice, Mabel 25	"
Rice, Ernest 25	"
Rimmell, H. H. 26	Hoytville
Randall, Allen 18	Hoytville
Parker, John 11	"
Randall, Chas. 8	"
Root, Wm. 34	"
Randall, William 16	"
Randall, Chas. 12	Mulliken
Randall, Frank 7	Hoytville
Randall, W. 7	"
Rimmel, Geo. 21	"
Reed, D. W. 11	"
Reed, Riley 4	Mulliken
Reynolds, N. Estate of 4	"
Reynolds, N. H. 1	"
Roland, Orin 18	Grand Ledge
Skipple, Mrs. C. 10	Mulliken
Sowles, Mrs. May 4	"
Sowles, Bert 8	"
Sterns, Henry 10	"
Sterns, M. 3	"
Spencer, Henry 8	"
Stacks, Northrop 2	"
Smith, Harvey 3	"
Smith, Frank 3	"
Spore, Jacob 11	"
Starkweather, R. 14	Mulliken
Sturdivant, Estate of 21	"
Sayock, Elmer 6	"
Sayock, Thomas 6	"
Sowles, Norman 5	"
Sanborn, R. F. 6	"
Stoler, David 3	"
Southworth, Geo. 3	"
Strong, C. W. 21	Hoytville
Stevens, Nancy 29	"
Stall, Jessie 10	"
Savage, Abram 15	"
Savage, A. 3	"
Snyder, James 29	"
Sevalt, Geo. 22	"
Stretch, Hattie 13	"
Spaulding, Mary 9	"
Spaulding, William 16	"
Spaulding, John 18	"
Shaner, Fred 15	"
Sayer, John 8	"
Sheets, John 26	Roxana
Sullivan, O. W. 36	"
Starkweather Owen 25	"
Thompson, Sammie 2	Mulliken
Triphagen, Geo. 4	"
Trim, Mrs. Terry 8	"
Thomas, Lewis 8	"
Thomas, Nettie 9	"
Triphagen, John 2	"
Triphagen, S. V. 11	"
Tobin, D. J. 20	Hoytville
Thomas, Sanford 21	"
Taylor, James 22	"
Taylor, Mrs. L. 3	"
Tobin, John 21	"
Vanoortwick, J. J. 29	"
Vanoortwick, Geo. 16	"
Vale, Laura 21	"
Vader, Jerry 22	"
Vanoortwick, C. 15	"
Vanhouten, John 16	"
Vanhouten, F. R. 16	"
Vickery, Thomas 27	"
Wellington, C. 8	Mulliken
Wright, Henry 1	"
Wilcox, L. H. 22	"
Wells, Mrs. 26	Roxana
Woolcut, W. 26	"
Wells, Mary 26	"
Waterman, Mrs. M. 35	"
Wilcut, J. J. 19	Hoytville
Warren, Geo. 10	Roxana
West, Frank 15	Hoytville
Whelpley, F. A. 8	"
Wheeler, Jerome 29	"
Wheeler, Hiram 29	"
Wanner, William 20	"
Williams, James 21	"
Wilcut, Jesse 19	"
Woolpert, Caleb 13	"
Woolpert, Fred 21	"
Wakely, Nathan 26	"
Whelpley, R. L. 14	Mulliken
William, Martha 6	Sunfield
Wise, Anna 21	Bismark
Wilcox, L. H. 3	Mulliken
Whelpley, H. L. 3	"
Whelpley, Ida 8	"
Wing, James 4	"
Watkins, William 8	"
Williams, James 3	"
Webster, C. A. 11	"
Wellman, C. H. 1	"
Williams, E. 11	"
Webster, Chas. 8	"

VILLAGE OF VERMONTVILLE.

Ambrose, C. M.
Ambrose, F. P.
Allen, D. W.
Allen, Spencer G.
Alderman, Lydia
Alderman, A.
Andrews, L. V.
Bowles, John
Brooks, Lewis
Boardman, Ed C.
Barrett, Hugh
Barber, Homer G.
Barber, M. D.
Barber, Daniel
Bailey, Frank J.
Benedict, William H.
Benedict, Fred H.
Benedict, Asa
Benedict, Geo.
Bronson, Myron
Benningham, John M.
Bodine, Mrs. Susan
Briggs, Ellen
Briggs, Alfred N.
Barnard, David
Brown, Sarah A.
Benton, Geo. A.
Chilson, N. C.
Cunningham, R. J.
Cross, Mrs. Ellen
Cross, Henry L.
Cross, N. D.
Cross, Arthur
Cross, Herbert G.
Collins, W. D.
Collins, C. C.
Cherry, Chas.
Campbell, N.
Campbell, Thomas
Curtis, Mrs. M. H.
Curtis, H. L.
Dunlap, Mrs. L. A.
Dickinson, Mrs. W. H.
Dickinson, H. S.
Dickinson, Naomi
Downing, J. C.
Dunsmore, M. W.
Durman, Chas.
Fleming, James
Fay, O. P.
Finley, D. R.
Fleming, James
Finley, Joseph
Fuller, N. A.
French, Mrs. E. A.
Gunn, Charlotte E.
Green, Dr. F. L.
Griswold, Mary B.
Griswold, William S.
Griswold, Isaac H.
Gunn, Robert
Hawkins, Dan J.
Hawkins, A. G.
Hawkins, G. W.
Hawkins, L. B.
Hall, Frank B.
Hall, Chas.
Hall, Sylvester
Hall, H. W.
Hall, Mrs. Ellen
Hall, Chas. D.
Hickey & Maher
Hickey, William
Imes, John
Imes, W. E.
Irving, H. B.
Johnson, William G.
Jones, Mrs. V. B.
Jordan, Mrs. A. L.
Kelley, Frank W.
Kelley, A. J.
Kelley, Barney
Knowles, F. A.
Kidder, Amos
Knapp, Mrs. Ezra
Kroger, C. J.
Loomis, Mrs. S. P.
Lake, A. D.
Lentz, Mrs. Fred
Lackey, J. A.
Lee, Mrs. H. W.
Lamb, Geo. J.
McCotter, Mrs. Simeon
Morey, G. H.
McConnell, Geo.
McConnell, Olvier
Mead, O. G.
Martin, Mrs. M. J.
Martin, H. J.
Mears, Eugene
Marble, Calvin
Norris, W. E.
Newton, H. M.
Pullman, Leroy
Pullman, Diana
Parmenter, Dr. William
Porter, Mrs. Cornelia
Porter, E. W.
Pierce, Mrs. James
Park, Chas.
Parks, M. H.
Paine, Mrs. F. A.
Paine, A. B.
Pendill, Fred
Pendill, Ellen
Purchis, D. M.
Purchis, L. A.
Paul, William A.
Peterson, Mrs. L. A.
Rawson, Edgar
Rawson & Lake
Randall, A. P.
Rhodes, John L.
Rickle, Joseph
Smith, Harley

VILLAGE OF VERMONTVILLE.

Sprague, Argelus	
Stone, J. B.	
Smith, B. J.	
Stiles, M. L.	
Scott, Mrs. Christena	
Snell, Dr. C. S.	
Stout, William	
Stebbins, O. O.	
Stebbins, Mrs. P.	
Sackett, James	
Sackett, William	
Sherman, A. M.	
Sherman, J. C. & Son	
Sheriff, Frank	
Satterlee, J. H.	
Tabbs, R. F.	
Tracey, Mrs. R.	
Tyler, Edwin	
Whitehearse, Samuel	
Walworth, Mrs. H.	
Wood, J. J.	
Williams, J. H.	
Whaling, Eugene	
Watson, James T.	
Vock, John B.	
Yates, N. A.	
Young, David	
Zemke, Herman	

VERMONTVILLE TOWNSHIP.

Averill, Edward 21	Vermontville
Bissell, A. B. 18	"
Benedict, Asa 19	"
Baker, Seymour 19	"
Barber, E. W. 6	"
Brigham, E. W. 19	Nashville
Brigham, Allie 19	"
Briggs, James 34	Vermontville
Benedict, Lihana 36	"
Bell, Montgomery 36	"
Barrett, Margaret 15	"
Beck, Joseph 23	"
Boehler, Albert 13	"
Boyer, Adam 14	"
Barnes, N. B. 23	"
Brown, H. H. 5	"
Baker, E. J. 5	"
Beck, Joseph 23	"
Baker, Wm. C. 33	Chester
Briggs, Ira 35	"
Brown, F. C. 18	Nashville
Cupp, Asa T. 29	Vermontville
Chaney, Robert 17	"
Chaney, Eliza 18	"
Compton, C. I. 30	"
Childs, Jas. H. 17	"
Carey, Eugene 25	"
Carey, Geo. W. 1	"
Campbell, Wm. 10	"
Campbell, N. 10	"
Campbell, E. 4	"
Cole, A. K. 11	"
Corey, Sidney B. 15	"
Childs, Wm. 4	"
Childs, J. D. 4	"
Childs, John J. 4	"
Cross, Joseph 23	"
Curtis, Leonard 33	"
Carter, John 31	Nashville
Darrow, Russell 30	"
Downing, Manly 16	"
Denton, A. P. 19	"
Deer, Mrs. Joseph 24	Vermontville
Davis, Walter 29	"
Davis, Mrs. Jonas 29	"
Dwight, Geo. C. 13	"
Dibeman, Halleck 1	"
Diehl, Jeremiah 2	"
Dunbar, Geo. W. 2	"
Dwight, C. H. 11	"
Dickinson, H. N. 17	"
Dooling, P. 5	"
Dooling, Jerry 5	"
Dooling, Andrew 5	"
Davis, John 36	Chester
Ellsworth, J. B. 3	Vermontville
Ettel, John 33	"
Fare, Chas. 35	"
Fare, Walter A. 35	"
Foote, Clark 33	"
Fuller, C. V. 24	"
French, Warren 31	"
Fowler, Mrs. C. 1	"
Farrah, Daniel 2	"
Fox, Perry 18	"
Falconer, Eleanor 18	"
Fishbaugh, R. P. 19	"
Ferris, William 7	"
Fuller, C. F. 23	"
Faust, Barna 35	"
Fremaire, Harrison 4	"
Grant, Hiram 30	"
Gates, Austin 31	"
Gearhart, John N. 19	"
Griner, P. N. 27	"
Green, J. L. 13	"
Green, A. O. 19	"
Gardiner, John 18	"
Grube, F. W. 6	Nashville
Granger, Geo. 33	Chester
Gunn, Orlando 14	Vermontville
Hawkins, Harvey 3	"
Hawkins, Jay 3	"
Hay, Frank 19	Vermontville
Hay, Adam 19	"
Hyde, Mrs. H. W.	"
Hollenbeck, C. C. 28	"
Hollenbeck, C. A. 33	"
Hawkins, Daniel 21	"
Hosvick, M. J. 23	"
Hawkins, Horace 16	"
Hauer, Henry 15	"
Hauer, Rufus 12	"
Harrenne, Geo. B. 16	"
House, Nelson 35	"
Hickery, Joseph 6	"
Hickery, Daniel 6	"
Hickery, Richard 5	"
Hickery, William 7	"
Hager, Herman C. 9	"
Hager, Sylvia J. 9	"
Haight, Fannie 8	"
Haight, Albert 8	"
Howell, Amanda 21	Nashville
House, Chauncey 31	"
Ives, W. E. 20	Vermontville
Jarrett, Ezra 11	"
Kroger, C. J. 29	"
Kibler, E. L. 1	"
Kirby, Robert 27	"
Kirby, Reuben 28	"
Kenworthy, Geo. 9	"
Leitterdy, Talma 30	Nashville
Lake, B. B. 11	Vermontville
Lake, John A. 22	"
Lake, Alonzo 2	"
Lynch, M. 7	"
Lute, John 24	"
Loomis, B. A. H. 4	"
Lemmon, A. B. 25	"
McGee, S. A. 11	"
Marks, John 1	"
McIntyre, John R. 25	"
Mears, Eugene W. 27	"
Moore, Chas.	"
Moore, Ed.	"
Maher, Michael 7	"
Maher, Mrs. John 8	"
Mowrey, Jacob 5	"
Muir, Robert 23	"
Marsh, Levi 34	"
Murrhouse, W. 8	"
Mull, David 10	"
Merritt, Mrs. Warren 3	"
Merritt, John 3	"
Mattson, W. D. 31	Nashville
Mix, Joseph 31	"
McDowell, Chas. 1	Vermontville
McLaughlin, David 36	Chester
Norton, I. C. 19	Vermontville
Newmond, Andrew 33	"
Nagle, Walter 28	"
Parker, W. J. 27	"
Perkins, M. D. 18	"
Purchis, D. M. 20	"
Purchis, Mrs. L. A. 19	"
Prudill, J. R. 43	"
Pohantus, Edward 15	"
Patterson, Lee 3	"
Patterson, Hollis Y. 4	"
Pember, Charlie 5	"
Palmateer, Joseph 18	"
Quinter, Thomas 6	"
Quinner, Geo. 29	"
Rawson, John 4	"
Rawson, E. W. 5	"
Rawson, B. F. 5	"
Remalie, John 10	"
Reynolds, Silas 34	"
Riley, Thomas 19	"
Rogers, Susan 17	"
Russell, Hiram 30	"
Rawson, Zera 16	"
Rawson, Fred E. 15	"
Remalie, Jesse 13	"
Snyder, A. S. 7	"
Surine, Horace 6	"
Stevens, John V. 21	"
Sprague, Ernest 11	"
Sibbrel, Edward 23	"
Sprague, Reuben 10	"
Stevens, G. B. 3	"
Scott, Geo. H. 16	"
Sprague, Mason C. 12	"
Sprague, E. J. 14	"
Sprague, Wm. H. 12	"
Spitler, Ed 1	"
Stowe, E. J. 18	"
Sprague, Fred P. 10	"
Sulsbey, Wm. 9	"
Smith, Artemus 19	"
Shepherd, Samuel 15	"
Surine, Ernest 7	"
Surine, W. E. 7	"
Schauder, John 4	"
Surine, Alfred 17	Nashville
Sline, Wm. M. 32	Nashville
Smith, Levi S. 31	"
Sprague, F. 22	"
Sherman, E. E. 23	"
Schram, Warren 32	Vermontville
Town, F. W. 2	"
Tassell, Wm. A. 29	"
Taylor, James 12	"
Trim, Austin 1	"
Tobin, John 1	"
Thayer, D. L. 28	Vermontville
Thompson, W. 21	"
Viele, Chas. 11	"
VanVleet, William 3	"
Wells, Ed. M. 3	"
Waite, W. A. 31	Nashville
Williams, A. R. 30	"
Williams, E. D. 30	"
Wyble, William 36	Chester
Well, William A. 1	Vermontville
Well, Fred L. 5	"
Weaver, Eugene 15	"
Wetherbee, R. L. 1	"
Wells, Perry H. 35	"
Wells, Clark E. 34	"
Walker, William 6	"
Ward, Mrs. Lockwood 22	"
Zemke, Mrs. Mary 29	"
Young, William 5	"

SUNFIELD TOWNSHIP.

Allen, Robert 21	Sunfield
Allen, Wm. 16	"
Allen, Fred 21	Bismark
Aungst, Daniel 2	Sunfield
Ballou, Abner 31	Vermontville
Bower, Stephen 31	Dellwood
Baker, J. C. 21	"
Bishop, Chas. 21	Sunfield
Bishop, A. B. 18	"
Bishop, Wm. 14	"
Bennett, Amos 28	"
Bishop, Delos 13, 22 & 24	"
Bosworth, Mrs. J. E. 35	"
Barnum, Wm. 24	"
Barnum, Chas. 21	"
Bosworth, Fred 35	"
Barden, Allen 36	Bismark
Bassore, E. M. 36	"
Barnes, D. B. 28	"
Bark, Aaron 33	"
Barnum, Wash. 28 & 25	"
Bole, Gideon, 25 & 26	"
Bennett, Wm. 31	Sunfield
Bower, Stephen 30	Dellwood
Bennett, C. A. 30	"
Bennett, Daniel 16	Woodbury
Baker, Elmer 30	Dellwood
Barton, James	Sunfield
Bottomley, Leo	"
Bera, E. D.	"
Bera, Warner	"
Bera, J. H.	"
Barnes, Mrs. O. L. 11	"
Berman, J. D.	"
Bliss, S. P. 2	"
Boyer, Wm. 5	"
Burns, Tompson	"
Bale, Wm. 12	"
Burns, Mrs. A. D.	"
Beecher, W. F.	"
Brown, C. H. 6	Woodbury
Brown, Clarence 6	"
Bateman, B. H. 6	"
Bennett, J. D. 15	Sunfield
Babel, Susan	"
Barden, Allen 33	"
Barden, John 28	"
Baughman, H. H. 22	"
Benedict, E. 27	Bismark
Bolles, James 32	Dellwood
Brown, S. H. 17-18	Sunfield
Brown, Chas. 21	"
Babcock, H. 27	Woodbury
Bishop, John 14	Sunfield
Bradley, Oliver 11	"
Beal, John 11	"
Bosworth, Mrs. O. S. 36	Bismark
Bosworth, Fred 36 & 35	"
Bosworth, Frank 36	"
Bennett 18	Woodbury
Beckhorn, Daniel 9	Sunfield
Barnes, M. 13	"
Briggs, John	"
Barnes, Hiram	"
Bowers, Maggie 11	"
Childs, J. A.	"
Collier, Fred 8	"
Collier, Edd	"
Collier, Chas. 8	"
Cogswell, Frank 8	"
Childs, Joseph 3-4	"
Collier, Geo. 30	"
Cheetham, Geo.	Sunfield
Collins, John 31	Dellwood
Clark, Floyd 27	Bismark
Cramer, Mrs. C. 23	"
Chipps, Wm. 21	"
Core, James 1	Sunfield
Crane, J. W. 14	"
Core, Amos 1	"
Carey, J. W.	"
Cheal, Jim 11	"
Carpenter, Chas. 14	"
Cornstalm, G. H. 12-21	"
Childs, C. S. 23	"
Chaney, T. H. 14	"
Chamlin, T. S.	Woodbury
Cooley, Mrs. Wm. 7	"
Clark, Elmer	"
Curtis, W. D. 33	Sunfield
Curtis, Webster 26	"
Cain, Abram 16-35	Bismark
Chatfield Peter 35	"
Carter, Mrs. 36	Bismark
Collins, A. D. 35	"
Cook, Frank 32	Vermontville
Conrad, Geo. 2	Woodbury
Chatfield, Henry 31	Vermontville
Curtis, W. D. 34	Sunfield
Cobb, H. B. 7	"
Clark, F.	"
Cone, Wm. O. 13	"
Coy, Henry 36	Bismark
Dund, S. C. 18	Sunfield
Dell, Adam 7	Woodbury
Daw, J. C. 34	Hoytville
Dilley, Robert 12	Sunfield
Davis, Margaret 3	"
Dunbar, John	"
Dralaman, D. H. 9	"
Day, J. E. 11	"
Darrow, J. H. 11	"
Dickinson, A. L. 35	Bismark
Davis, Wm. 22	Sunfield
Davis, A. J. 13	"
Dickinson Geo. 20	"
Downs, Amos 21	"
Downs, R. M. 16	"
Dewel, Jerome 27	Bismark
Downing, A. J. 16	"
Dewel, Orlando 27	"
Dewel, Nancy 27	"
Eckard, Wm. 10	Woodbury
Eldred, John S.	Sunfield
Edwins, Wm. 8	Woodbury
Fremyre, W. L. 34	Bismark
Flewelling, Mrs. J. 33	"
Fay, W. C. 19-20-30	Dellwood
Fast, J. A. 30	"
Fenshaw, Louis	Woodbury
Fender, Van 28	Sunfield
Frantz, Nichols 21	"
Fremyer, W. L. 34	Vermontville
Frantz, Benj. 34	Bismark
Frantz, Chris 16	Sunfield
Frantz, Mose 16	"
Fiser, Alex. 12	"
Fleetham, Joseph 1	"
Fleetham, Edd. 1	"
Day, J. E. 11	"
Figg, Henry 12	"
Fender, Pete 11	"
Figg, Stephen 1	"
Freese, Geo.	"
Fields, A. J. 9	Sunfield
Frantz, Wm. 23	"
Frantz, Frank	"
Fryfogle, H. 15	"
Fritz, Wm. 19	"
Fisbel, J. L. 21	"
Freeman, Wm. 30	"
Fremyre Bros. 34	Vermontville
Guy, Hezekiah 19 & 30	Woodbury
Grant, Isaac 30	Dellwood
Garringer, J. S. 30	"
Gorham, Dan 21	Bismark
Garringer, E. L. 30	Dellwood
Gow, D. M. 9	Sunfield
Gelden, J. P. 12	"
Gross, Henry 9	Woodbury
Gregg, Geo. 19	Sunfield
Gorham, Wm. 21	Bismark
Gleason, Amanda 30	"
Gray, James 5	Woodbury
Gray, Mart 6	"
Gidy, Jessie 8	"
Hodgkins, M. S.	Sunfield
Hammond, J. H.	"
Holst Bros.	"
Hart, Henry	"
Hartwell, Ira	"
Hoover, John 22	"
Hyde, John 13	"
Hough, Geo.	"
Hiddinger, G. V. 7	Woodbury
Hiddinger, Jacob	"
Hunt, Edwin 5	"
Haskins, C. M. 6	"
Hager, Milton 4	Sunfield
Hager, Joseph 9	"
Hale, Sim 18	Woodbury
Hough, Geo. 18	Bismark
Hawkins, H.	Sunfield
Hammond, W. H. 22	Dellwood
Hill, Wm. 19	Woodbury
Head, Geo. 31	Vermontville
Harvey, John 17	Woodbury
Hitt, Geo. 30	Woodland
Hager, Orson 31	Dellwood
Hecker, Chri. 30	Vermontville
Hammond, Elmer 22	"
Hammond, Marques 27	Bozzana
Hicks, John 26	Bismark
Hunter, Thos. 33 & 28	Dellwood
Hammond, Wm. H. 12	"
Harris, Martin 29	"
Hager, D. A. 24	Sunfield
Hager, Wm. 16-17	"
Hunter, H. S. 23	Dellwood
Hunter, J. R. 26	"
Hunter, Allie 21	"
Hunter, Mrs. Jas. 27	"
Hoida, Hannah 1	Sunfield
Hay, Geo. 6	Woodbury
Horn, Mrs. Jacob 6	"
Homes, W. S.	"
Holton, Geo. 27	Dellwood
Ives, Newton 16	Sunfield
Jackson, Chas.	"
James, Geo.	"
James, Jesse 6	Woodbury
Jackson, Chas. 12	Sunfield
Jackson, Geo. 13	"
Jones, W. A. 8	Woodbury
Jones, Mrs. 8	"
Jacobes, Leonard 21	Sunfield
Knapp, Eugene	"
Kent, P. M.	"
Kreacher, Geo.	"
Knapp, D. W.	"
Knapp, Henry	"
Knapp, Horace	"
Knapp, Mrs. Frank	"
Kramer, Mrs. A. 9	"
Kramer, A. 4	"
King, W. R.	"
Kenworthy, W. M. 33	Dellwood
Kinsle, E. H. 35	Bismark
Kidd, Frank 1	Sunfield
Kost, Wm. P. 3	"
Krebbs, Wesley 5 & 8	"
Kimmel, Amos 6	"
Lester, T. W. 29	"
Lyon, W. A. 32	"
Lamb, Geo. Jr. 20	"
Laird, Mrs. S. 24	"
Lovell, L. D. 31	Dellwood
Lowe, E. W. 26	Lake Odessa
Lusk, G. W. 1	Sunfield
Litchfield, D. 11	"
Leigh, S. M. 21	Dellwood
Legge & Jenkins	Sunfield
Lenhart, Hart	"
Lenhart, Frank	"
Lenhart, Otis	"
Lapo, J. H.	"
Lacey, Michael 8	"
Laird, Wm. 8	Woodbury
Lapo, Chas. 6	Woodbury
Laird, R. M. 6	"
Laird, Chas. 7	"
Lowe, E. W. 26-35	Bismark
Lemmon, L. G.	"
Lemmon, J. M. L. 36	"
Leigh, F. B. 5	Woodbury
Leigh, C. W. 5	"
Lombard, Mrs. 6	"
Mallory, Mrs. J. D. 19	Sunfield
McIntyre, Ed 36	Bismark
Middaugh, Dee 16	Sunfield
Meyers, Daniel 11	"
Meyers, V. 8	"
Meyers, Hendly	"
Meyers, H. C.	"
McGlayre, Chas.	"
Magden, Frank 16	"
Magden, Henry 10	"
Mathews, T. C. 11	"
Mead, John 6	Woodbury
Morgan, Henry	Bismark
Moore, David 36	"
McNabb, J. 35	"
McWhorter, H. O. 24	Sunfield
Miller, Fred 7	"
Mason, Geo. 17	"
Moore, Mont 35	Sunfield
Middaugh, Martin 6	"
Munford, A. B.	"
Nichols, P. M.	"
Neff, H. A. & 10	"
Nead, John 7 & 9	Woodbury
Neff, Erwin 34	Bismark
Nead, Mary 9	Sunfield
Nead, James 9	"
Paris, Jessie 33	Dellwood
Preston, D. M. 21	Bismark
Pettry, Harper 27	Dellwood
Phillips, James 27	Bismark
Phillips, Ira 27	"
Perkins, Josiah 9-16	Sunfield
Perkins, S. W. 9-8	"
Perkins, N. 9	"
Pratt, Geo. 21	"
Phillips, Norm 28	"
Pratt, Thos. 29	"
Potter, Wm. H. 23	Bismark
Prescott, Andrew 27	"
Oster, Henry 31	Sunfield
Omer, H. P.	"
Pifer, Jacob 18	"
Peck, J. L. 1	"
Paris, Wm. 1	"
Park, John 9 & 2	"
Putnam, M. O. 4 & 9	"
Palmer, Christian 5	"
Putnam, C. M. 9	"
Petling, Randolph	Sunfield
Peabody, Fred	"
Peabody, J. H.	"
Peabody, Lemon	"
Peck, Mrs. M. A.	"
Palmer, John H.	"
Rawson, John 12	Vermontville
Richards, Harry	Sunfield
Repto, Henry 31	Woodbury
Rogers, J. H. 18-19	"
Roberts, Frank 19	Woodland
Rawigle, John 10	"
Raffler, John 14	"

SUNFIELD TOWNSHIP.

Rowson, A. M. 31	Bismark
Hullsen, Wm. 25	"
Rawson, John 27	Vermontville
Rawson, Frank 27	"
Reader, Ben 1	Sunfield
Richards, Clark 15	"
Richards, Mary 1	"
Reed, Lyman	"
Reed, Wm. 11	"
Rumfield, Mrs. K. F	"
Rawson, H. A. 14	"
Rosenfelter, Jacob 31	Vermontville
Rame, Mahlon	Sunfield
Roland, Alvina 17	Woodbury
Rapr, John 8	Sunfield
Roberts, Frank 12	"
Richards, Geo. 5-11	"
Richards, Frank	"
Richards, LeRoy	"
Rogleson, Wm.	"
Ramsey, Mrs. M.	"
Ramsey Bros.	"
Ramsey, W. J.	"
Richardson, Wm. R.	"
Ruff, Samuel 15	"
Seitz, Geo. 31	Bismark
Seltzer, M. 19	"
Swick, Geo. 31	Dellwood
Story, Rev. Wm. 31	Bismark
Sackett, S. M. 24	"
Sinclair, H. 31 25	"
Smith, David, 6	Woodbury
Smith, Geo. 6	"
Shetter, John 5	"
Shetter, John, Jr. 5	"
Shaver, Joseph 4	"
Stank, Daniel 4	Sunfield
Sigas, G. W. 5	Woodbury
Shaver, Wm. 11	Sunfield
Spencer, T. J. 9	"
Smock, John 12	"
Smith, Jacob 11	"
Stevens, Chas. 19	"
Shaffer, David 12	"
Sanford, Mrs. M. 13	"
Shamblin, T. 6	"
Slotsick, J. W. 6	"
Sprague, C. P. 8	"
Seaton, Wm. 15	"
Spragues, O. U.	Woodbury
Slinchcomb, M.	Sunfield
Snyder, J. L.	"
Smith, Sol. 27	Bismark
Stinchcomb, T. E	Sunfield
Stinchcomb, David	"
Snyder, C. S.	"
Smith, Mrs. B.	"
Stout, Henry	"
Snyder, K. W.	"
Stinchcomb, Joe	"
Shaver, Hugh	"
Sprague, Fred 10	"
Stratber, Jacob	"
Smith, Geo. 10	"
Sperry, L.	"
Shelter, J. R. 6	Woodbury
Shelter, Sylvester 8	"
Smith, Daniel 8	"
Smith, Mrs. J. 6	"
Scheel, Geo. 6	"
Shaffer, Cyrus 21	Sunfield
Sackett, N. 12	"
Smart, C. U. Est. 21	"
Sackett, Wm. 18	"
Sherrard, Jim	"
Spacer, H. R. 4	"
Trowbridge, Ira 15	Sunfield
Town, C. N. 29	"
Town, Mrs. P. A. 29	"
Tomlinson, J. 15 & 22	"
Teal, Henry 18	"
Thorp, Geo. 11 & 14	"
Town, J. N.	"
Torpy, W. J. 4	Woodbury
Trumner, Peter 8	"
Taylor, Edwin 11	Sunfield
Turner, T. P. 27 & 24	"
Trabster, Caleb 19	Dellwood
Trim, Amanda	Sunfield
Turner, W. R. 10	"
Thurston, Oscar	"
Tickner, Geo.	"
Truman, H. A. 15	"
Vanhouton, H. 6	Woodbury
Vanamie, W. E.	Sunfield
Vahouton, C. M. 24	"
Varney, J. H. 29	"
Van Buren, T. B. 28	"
Vandiers, Wm. 27	"
Velte, John 24	"
West, D. E. 1	"
Wager, M. B.	"
Wilson, John	"
Walsh, Tom	"
Wilson, L. O.	"
Weaver, Sam	"
Wright, Chas.	"
Wood, John	"
Woolcott, John	"
Wells, John 23	Bismark
Walsh Mrs. Thomas 24, 23 & 26	"
Welsh, P. J. 23	Sunfield
Wilson, Mrs. Sophia 28	Bismark
Wise, S. A. 30	Dellwood
Winters, Ella 18	Woodbury
Weiss, Schuyler 22	Dellwood
Wells, Wm. A. 23	Vermontville
Wright, Mary A. 30	Dellwood
Wright, Geo. B. 28	"
Wright, Mrs. J. D. 22	"
Wright, L. B. 20	"
Witherel, W. R. 14	Sunfield
Wescott, M. W. 22	"
Weeks, M. S. 22	"
Wonick, Geo. 9	"
Wells, W. B.	"
Welch, Rachel	"
Wilber, R. F.	"
Wells, Chas. 11	"
Welch, John 23	"
Winters, John 11	"
York, H. J.	"

VILLAGE OF OLIVET.

Allen, Mrs. S. P.
Barrows, John M. Estate of
Barrows, Giles J.
Beecher, Chas.
Blanchard, Mary A.
Blanchard, Geo.
Blanchard, Flora
Blanchard, N. S.
Bosworth, F. M.
Bracey, Guy
Brockman, F. W.
Burroughs, Susan
Carver, James A.
Chapman, Ed
Clark, Albert
Curtis, Chas. B.
Cook, Peter V.
Cone, Margaret J.
Corey, E. B.
Dermott, Clarence
Daniels, Prof. Joseph L.
Davis, Randolph
Davis, Henry B.
Deringer, Robert
Dowler, Mrs. Mary
Dickson, Prof. Chas.
Dewey, Mary
Darringer, Irwin
Ely, E. N.
Ellis, Prof. Geo. N.
Emmerson, William
Eastbrook, Mrs. Eva
Edwards, Jesse
Ely, Fred N.
Fuletts, Benj. Estate of
French, M. K.
Fuller, H.
Farlin, Wm.
Gage, Fritz H.
Gutsans, Geo.
Griffin, John
Getty, Pamela
Gilbert, Whitney
Goodwin, Henry Estate of
Goodrich, Darius
Green, Cynthia
Green, H. K.
Green, A. P.
Hadden, Smith
Halstead, Jonas A.
Hamer, M. A.
Hart, D. M.
Herrick, Henry
Harrington, Jas.
Hickok, Georgia K.
Hoes, Geo. W.
Hosford, Orsamel Estate of
Horn, Benj.
Herrick, Martha
Insley, William G.
Kedzie, Ella N.
Kevary, Champlin
Kayes, K. B.
Knapp, Prof. Geo. A.
Kellogg, Prof. J. L.
King, Prof. Hamilton
Knox, Henry E.
Knox, Fred L.
Knox, Daniel
Keeler, Hannah
Laughbaugh, Augusta
Lee, Albert L.
Lockett, Wm.
Mead, Dr. Chas. H.
Mack, Martin
Mapes, S. W.
Martin, Mrs. E. C.
Martin, I. E.
McDonald, Henry
Mills, Elvira L.
Montague, Chas.
Moore, Geo.
Montgomery, Prof. Stewart
Morton, Emma L.
McDonald, Elmer
Nettleton, Emeline
Nash, Polly
Norton, E. F.
Niles, Mrs. M. J.
Pinch, S. W.
Pinch, Fannie
Peters, Warren A.
Priters, Truman
Prosser, Thomas
Prosser, Mrs. L. D.
Remalia, Henry
Ramsey, Francis
Reed, F. L.
Reynolds, Juliet
Bradgee, O. J.
Shaffer, A. U.
Savage, Amos T.
Shaffer, Henry
Shoop, Anna
Sidaway, P. E.
Burr, Emily J.
Sperry, Pres. Willard G.
Storr, Frank
Storr, E. J.
Shore, A. K.
Stone, Carrie A.
Swint, J. K.
Toppling, Albert
Townsend, John
Tobin, Albert
VanGelson, Harriet
Warren, Dr. A. K.
Waterson, Edward
Waddell, Nancy S.
Wever, David
Warner, Geo. N.
Wood, Wallace S.
Weimore, Fay
Wise, Prof. Sam'l L.
Willis, Miss Anna

WALTON TOWNSHIP.

Andrews, John 27	Olivet
Andrews, Chas. 22	"
Allen, Harvey T. 28	Partello
Allen, Earl H. Est. 14 & 78	"
Hassett, James E. 18	Charlotte
Bryan, James D. 12	"
Bryan, Horton A. 12	"
Brown, Homer 18	"
Barber, Emma 13	"
Hatherfield, Orlo C. 14	"
Bowen, John J. 17	Olivet
Bowne, Geo. F. 17	"
Dagner, Orlo 21	"
Bennett, Lewis A. 22	"
Burroughs, E. G. 22	"
Buckland, Orville P. 11	"
Burroughs, May C. 23	"
Beecher, Harriet Est. 28	"
Butler, L. H. 23	"
Belding, A. A. 26 & 27	"
Burroughs, Myron 33	"
Bryant, S. G. 27	"
Baldwin, C. M. 27	"
Burroughs, F. W. 28	"
Bosworth, Alonzo 23 & 31	"
Bradley, Otis E. 7	"
Bradley, Merritt K. 8	Alnger
Bowen, Malinta 6	"
Brea, John 18	"
Bracey, Frank 8	"
Bradley, J. O. 5 & 8	"
Bradley, Ira D. 6	"
Cornell, Thomas 1	Charlotte
Cranell, John A. 8	Alnger
Cooper, James 8	"
Cole, Sarah S. 4	"
Crampton, Levi 24	Olivet
Colby, Dexter 21	"
Campbell, Geo. A. 8	"
Cornell, Lewis 13	"
Cornell, William 13	"
Curtis, Samuel 22	"
Garver, Andrew J. 34	"
Carpenter, Seth 25	"
Crampton, Oscar 15	"
Coon, Edwin 15	"
Claus, D. J. 20	"
Carver, Richard 27	"
Consett, Kara 28	"
Cornell, Aaron T. 14	"
Dillon, Otto 4	Charlotte
Day, Albert C. 7	Alnger
Day, Curtis A. 7	"
Darner, James 5	"
Dusenberry, Walter 6	"
Day, Sylvester Est. 6 & 7	"
Darby, William L. 25	Olivet
Davidson, Mrs. Dewitt 15	"
Donaldson, Frank 26	"
Deringer, Robert	"
Darby, Julius Est. 31	"
Edson, Andrew 9	Alnger
Evans, David 7	"
Haner, Caleb M. 4 & 5	"
Ely, E. B. 16 & 21	Olivet
Edson, Philip 3 & 10	"
Force, Amos 1	Charlotte
Fink, William H. 4	"
Frains, Joseph 4	Alnger
Fame, Henry Est. 9 & 16	Olivet
Foote, Henry 14	"
Foote, Mrs. A. E. 17	"
Finch, Willard 22	"
Francis, Lester P. Est. 21	"
Foote, Mrs. M. W. 9 & 16	"
Fisher, Frank A. 3	"
Fisher, N. N. 13 & 14	"
Gamble, Ambrose 9	Alnger
Gorman, John, Sr. 18	"
Gorman, John, Jr. 8	Alnger
Hadley, Thomas 7	"
Gamble, Alex 1	"
Griffin, Orrin 3 & 9	Olivet
Griffin, Ed. 9	"
Galusha, Warren 27	"
Green, William 31	"
Hobart, Jane 21	"
Hatch, Giles 13	"
Haddon, Geo. H. 11	"
Hattis, William 15	"
Hattis, Henry 14	"
Harris, Thomas 22	"
Hogle, Fayette 18 & 19	"
Hoffnagle, William C. 21	"
Harrington, William A. 17	"
Hall, C. M. 28	"
Hart, William 34	"
Hackman, L. E. 27	"
Harris, John 28	"
Hayes, Peter S. 4	Alnger
Hoskins, Merritt 18	"
Hass, G. E. 6	"
Hart, Jesse E. 1	Charlotte
Hockenberry, Jay 2 & 11	"
Hitchcock, Oscar 13 1	Olivet
Hockenberry, Martin 2 & 11	Charlotte
Hockenberry, Clinton, 11 & 12	"
Ives, Philetus 21	Olivet
Jewell, Lucy 24	"
Jewell, Mills 23	"
Johnson, Mrs. C. 22 & 28	"
Johnson, Mark L. 18	Charlotte
Johnson, S. R. 12 & 13	"
Kunkle, Jacob 10	"
Kunkle, Michael, Sr. 9	"
Kunkle, Michael, Jr. 3	"
Kunkle, Jacob C. 9	"
Kay, Edmund H. 21	Olivet
Kelly, Mary A. 23	"
Kelly, J. A. 24	"
Kay, Edmund M. 21	"
Kunkle, Jacob E. 9	Alnger
Lybeck, W. F. 7	"
Lowe, A. 18	Charlotte
Lyon, Mrs. C. C.	Olivet
Mahan, Henry 13	Charlotte
Mauyer, Joseph 1	"
Mobilla, Joseph 4	"
Myers, William 1	"
McCreery, Devillo 4 & 9	"
McCreery, Hiram 4 & 9	"
Martin, Frank 8	"
Miller, A. H. 10	"
Myers, Fred E. 4	"
Martin, Chas. 4 & 5	"
Miller, Mrs. Marie 15	"
Machin, Stephen Est. 4	Alnger
Machin, William 9	"
McWetey, F. H. 6	"
May, Hayles H. 5	"
Norman, Edwin H. 6	"
Messenger, Mrs. Orrilla 16	"
Murray, Harvey 26	Partello
Mattison, Spencer E. 23	Olivet
McDonald, Geo. 20	"
Mack, H. H. Est. 23	"
Miller, Alden 22	"
Messenger, David H. 15	"
Mark, Jacob Est. 16-20	"
Mott, Berger 23	"
Mott, James E. 23	"
Menker, H. W. 14	"
Marsh, C. May 25	"
Martin, E. B. 21	"
Marker, H. W. 14-15	"
McMertal, William 20	"
Mattison, Eugene E. 24-25	"
Marsh, Albert M. 9	"
Nye, Martillo A. 13	Charlotte
Nye, Ed. E. 18	"
Nye, Earl O. 18	"
Newland, Mrs. A. N. 22	Olivet
Newcomb, Carrilo A. 4	Alnger
Ogden, William Estate of 9	"
Oakes, Orion 16	Olivet
Ochenbein, Meinrad 4	Charlotte
Perry Harwick 7	Alnger
Perry, Orrin F. 7	"
Pierce, Loraine 8	"
Prosser, William 25	Olivet
Palton, L. C. 17	"
Pelton, Winfield 31	Charlotte
Robinson, Maryette 18-23-26	Bellevue
Roberts, T. A. 4	Alnger
Rolfe, Drusie 7	"
Roberts, William C. 5	"
Roberts, Homer A. 10 & 11	Charlotte
Roberts, Isaac N. 10 & 11	"
Ratter, Dan 7	"
Raidel, J. W. Estate of 3	"
Reed, Geo. A. 24	Olivet
Raynor, Edson W. 30	"
Richards, William E. Est. of 23	"
Rogers, Wm. 24 & 33	"
Raynor, K. S. 30	"
Brassuer, Jacob 35	"
Reynolds, J. W. 16	"
Smith, Geo. W. 3	Charlotte
Sharer, Chauncey 12	"
Smith, Albert C. 1	"
Starkweather, Geo. W. 1	"
Sheldon, David F. 2	"
Sherman, Eliza M. 21	Olivet
Scott, Wm. 24	"
Stamp, Isiah 10	"
Swift, F. M. 22	"
Scott, C. E. 18 & 24	"
Stone, J. A. Estate of 27	"
Salisbury, Ezra 21	"
Stuart, John P. 23	"
Stiles, Samuel J. 9	"
Scott, Earl B. 26	"
Storr, Emily J. 27	"
Storr, Fred 27	"
Salisbury, Orpha 24	"
Sexton, Weston M. 20	"
Stoddard, Bart 28	"
Sweet, Geo. W. 21	"
Scott, Elizabeth 24	"
Stalls, Frank 15	Alnger
Sessions, Hiram 22	"
Savery, John 1	"
Thornton, Lucian 8	"
Taggart, Lester H. 7	"
Thorn, John 22 & 27	Olivet
Taylor, Franklin W. 24	"
Tillotson, Mrs. 23	"
Valland, Herman 11	Charlotte
Wight, Wellington 5 & 6	Alnger
Wightman, Allen S. 4 & 5	"
Wight, Phebe A. 3	"
Woodbury, Chas. 5	"
Wightman, J. M. Estate of 7	"
Wickham, N. O. 1	Charlotte
Wells, Levi 18	"
Weeks, Dart	"
Wilson, Chas. 21	Olivet
Walker, Alex W. 23	"
Willis, Sumner 27	"
Wyble, Rachel 10	"
Waterson, Edward 26	"
Willoughby, Chas. 21	"
Waterson, Thomas 35	"
Walsh, Patrick 14 & 15	"
Wagoner, Mrs. Maria 16	"
Weeks, Elmer 10	"
Wagoner, Freeman 15	"
Wagoner, Isaac Estate of 16	"
Whitbeck, Richard E. 14	"
Wyble, William 23	Partello

WINDSOR TOWNSHIP.

Allen, Geo. M. 17	West Windsor
Allen, N. B. Est. 17	"
Ashley, Jabez 21	Dimondale
Armstrong, Mrs. E. G.	"
Andrews, G. O.	"
Almond, Thomas 8	"
Armstrong, M. 19	"
Ashley, Merritt 29	"
Brigham, B. F.	Lansing
Butler, L. D. 16	"
Boera, F. 16	"
Baker, A. C. 15	"
Bishop, Jacob 8	Charlotte
Bell, R. J. 18	Eaton Rapids
Baxter, Elias 26	"
Barnes, L. L. 7	Potterville
Buchholtz, Albert 6	"
Baldwin, Geo. 20	"
Brandeberry, Isaac 19	West Windsor
Boss, Mrs. Carrie 7	"
Bateman, N. P. 25	Dimondale
Batley, Geo.	"
Bramble, Mrs. Caddie	"
Bement, Loren 16	"
Beem, S. J.	"
Berner, A.	"
Berner, A., Jr.	"
Bramble, Chas.	"
Barge, G. S.	"
Bell, James A.	"
Bryan, Daniel 25	"
Bryan, J. B. 26	"
Bishop, E. G. Est. 26-27	"
Bailey, M. H. 26	"
Barber, G. A. 25	"
Barber, O. M. 22	"
Bishop, John 19	"
Billman, Emma 3	"
Barnett, Robert 29	"
Bigrow, Samuel 20	"
Bailey, Will H. 21	"
Beckwith, E. G. 24	"
Bohannon, Marie	"
Bohannon, Reed	"
Bates, Frank 22	"
Beal, John 13	"
Buckingham, Mrs. M. 11	"
Brosseling, B. Est.	"
Barrett, M. H.	"
Corwin, H. J. 23-24	"
Croy, T. J. 10	"
Cook, A 12	"
Carlton, H. S. 25	"
Courter, J. C. 22	"
Cogswell, Clark 11	"
Carlton, A. D. 21	"
Canedy, Fred 22	"
Carlton, H. S. 22	"
Crisher, Edwin	"
Cogswell, James 22	"
Carpenter, D. G. 19	"
Carrington, A. W. 9	"
Carpenter, H. 9	"
Crall, S. E.	"

WINDSOR TOWNSHIP.	
Carlton, A. D. 35	Dimondale
Cole, Allen 18	''
Crane, W. H.	''
Commerworth, Frank 26	''
Clark, A. 1	Lansing
Clark, Mrs. Elias 4	''
Croy, Henry 7	Potterville
Carman, Richard 19	''
Creasy, Samuel 34	
Covey, H. H. 31	''
Carman, H. M. 19	West Windsor
Christian, Ida M. 19	''
Christian, Mary 19	''
Capwell, Mrs. H. 11	''
Cook, Harriett 7	''
Canedy, Wm. 26	Eaton Rapids
Clark, Joseph 35	''
Catlin, Henry 29	''
Depue, F. E. 27	Dimondale
Deane, Irvey 7	''
Disbrow, Mrs. S.	''
Disbrow, M. F.	''
Drumm, H. 13	''
Dorr, Solomon 4	''
DeWaters, David 24	''
Delta, N. 8	''
Drew, John 6	West Windsor
Ellis, Mrs. M. R. 7	''
French, John R. 6	Dimondale
French, A. R.	''
Flagler, Chas. P.	''
Fish, J. W. 10	''
Fry, Chas. W. 5	''
Fry, Ellis 9	''
Finsloss, Joseph 2	''
Finsloss, L. 10	''
Flinchbaugh, E. A.	''
Fish, John W. 15	''
Ferris, Chas. 34	''
French, Silas 28	''
Fancher, Joseph 34	Kingsland
Freyre, Frank 33	Potterville
Gilbert, H.	Dimondale
Gundy, M. A.	''
Gilbert, Geo.	''
Gunderman, Rev.	''
Gregory, J. B.	''
Gardner, C. A.	''
Gibbs, John P. 24	''
Gibbs, Walter 21	''
Gilbert, C. B. 15	''
Gardner, W. D. 1	''
Gladden, A. B. 4	''
Gentner, John 8	''
Ge'schafer, Wm. 28	''
Garlan, Gideon 33	Eaton Rapids
Gelsenhafer, G. 19	Potterville
Getter, Mrs. Mary E. 31	''
Hilliard, Lee 21	Dimondale
Hilliard, Mark 21	''
Hetrick, John 18	''
Holcomb, Chas. 23	''
Holmes, John Q. 18	''
Herr, Jacob 19	''
Hetrick, Samuel 11	''
Hathaway, W. 1	''
Hethorn, Frank 25	''
Hemlinger, B. 28	''
Holloway, Ensign 23	''
Hunt, E. W.	''
Hrook, Alex 9	''
Haner, Seymour 5	''
Hall, N. P.	''
Hall, Tyler	''
Hull, Whitman 22	''
Holt, Eli 21	''
Hess, Isaac	''
Hough, Samuel 15	''
Hyde, William J.	''
Huntington, Mrs. J. Z. 9	''
Harris, E. C. 1	''
Hall, Joseph 19	''
Hall, John E.	''
Hotz, Chas.	''
Hurlburt, W. T. 12	''
Horton, N. B.	''
Hull, Miles 21	''
Hall, Mrs. S. J. 28	''
Hayter, Samuel 3	''
Hethorn, William 25	''
Howard, Moses 31	''
Hoffman, Russell 4	''
Howard, Eugene	''
Howard, J. D. 22	''
Holmes, A. D.	Lansing
Huber, Mrs. H. 24	Grand Ledge
Harmon, Fred 6	West Windsor
Hunter, Samuel 6	''
Henderson, Mrs. S. W. 18	''
Hough, John 7	''
Hamilton, Frank 1	Millett
Inselman, T. A. 18	West Windsor
Jaenicke, Henry 8	Dimondale
James, O. B. 11	''
Jones, J. J. 27	''
Jewells, William	''
Jones, H. C. 25	''
Jones, Orlando 20	''
Kolb, Samuel 12	''
Koon, Mrs. Mary 21	''
Kimball, Frank 33	''
Kelley, Leroy 29	''
Kook, Theodore	Dimondale
King, John M. 2	''
Kiper, Anthony 27	''
King, Milton 11	Lansing
Koon, Addison 23	Eaton Rapids
King, Mark 21	Potterville
Keller, Mrs. Jacob 1	Lansing
Lonsberry, W. E. 12	Dimondale
Lonsberry, James 1	''
Lemon, William 1	''
Longmate, Enoch 12	''
Leisenring, J. B. 22-32	''
Leisenring, Oscar 27	''
Lamphere, T. C. 1	''
Lamphere, H. N. 1	''
Lapack, H. A. 2	''
Leisenring, K. B. 21	''
Leisenring, Gideon 21	''
Longmate, K.	''
Lane, A. J. 26	''
Lewis, Chas. H. 34	''
Lockwood, James 34	Kingsland
Little, William 33	Potterville
Lee, John 21	''
Little, Daniel 20	''
Loomis, Mrs. M. A. 18	West Windsor
Lewis, Mrs. Orla 17	''
Lewis, J. L. 17	''
McCllintock, Geo. 15	Dimondale
McKim, Mrs. S. L.	''
Morrison, F. B.	''
Morrison, C. A.	''
Merritt, F. A.	''
Murphy, Mrs.	''
Murphy, Homer	''
Minis, Walter	''
Monroe, Julia	''
Mathews, Samuel 18	''
Miller, William 15	''
Murphy, John 15	''
McClintock, Geo. 10	''
McCarty, Tim 1	''
Murray, William 10	''
Marks, Emma 22	''
Morrison, K. F. 22	''
Miller, J. F. 10	''
Moore, N. 2	''
Musler, G. W. Estate 20	''
Martin, Samuel 20	''
Mengre, Wesley 31	Charlotte
Moorer, Frank 31	Potterville
Norris, Thomas Estate 23	Dimondale
Norris, Francis 5	Lansing
Norton, C. K.	Dimondale
Oliver, Alex	''
Oxendale, Isaac 33	Kingsland
Parsons, N. 17	Dimondale
Pray, Mark 27	''
Pray, Geo. D. Estate 29	''
Pray, F. G. 32	''
Pray, Andrew 21	''
Parsons, D	''
Peck, Mrs. S. M.	''
Potter, Cass. 9 & 16	''
Porter, A. F. 10	''
Pierson, J. W.	''
Pierson, M. M.	''
Porter, Mrs. A. F. 16	''
Phinney, Frank	''
Phinney, Fred	''
Partello, Bishop 10	West Windsor
Pinch, J. O. 7	''
Perkins, Geo. 1	Millett
Paine, Rowland 4	''
Peck, Lucius 9	West Windsor
Peck, G. F. 1	''
Proctor, A. G. 23	Potterville
Potter, J. J. 21	Kingsland
Robinson, C. J. 6	Dimondale
Robinson, Jerome 5	''
Roe, Daniel	''
Haas, Mrs. S.	''
Rose, Mrs. P.	''
Rankin, John	''
Rose, Geo. W. 21	''
Rogers, Mrs. Anna	''
Robbins, John 15	''
Rinkle, Henry 12	''
Rinkle, G. W. 13	''
Reeves, Esther 27	''
Reigle, Jacob 29	''
Rosell, Emmor 34	''
Reddick, N. 26	''
Reynolds, Geo. 2	''
Rogers, G. W. 31	Potterville
Root, Sylvanus 31	''
Reusch, Geo. 2	Millett
Reusch, Ellas 2	''
Roberts, Wm. C. 26	Eaton Rapids
Roberts, Elias 26	''
Redfield, Mrs. C. 22	''
Redfield, Geo. 22	''
Richardson, Frank 33	West Windsor
Root, E. C. 31	''
Randall, A. G. 6	''
Rogers, Wm. A. 7	''
Rose, Mrs. F. J. 17	''
Rumsey, S. F. 17	''
Schermerhorn, Mrs. L. 15	Dimondale
Selgford, L. W. 15	''
Spafford, Frank 15	''
Standish, Vol	''
Shively, Mrs. J.	''
Standish, Mrs. R. S. 15	Dimondale
Sage, S. F. R. 15	''
Schermerhorn, L.	''
Skinner, C. C. 22 & 21	''
Skinner, Julia 21	''
Spitler, J. M. 25	''
Searles, Louisa 35	''
Stanton, Otis 35	''
Stanton, A. G. 35	''
Smith, Isaac 4	''
Schmidt, Mrs. J. 2	''
Schmidt, John G. 10	''
See, Jay 9	''
Sienbel, A. W. 17	''
Streisel, Ernest 17	''
Sloan, T. M. 13	''
Schermerhorn, D.	''
Surphur, Mrs. H.	''
Smith, G.	''
Skinner, F. B.	''
Smith, Henry 19	''
Skinner, B. H. 13	''
Stiefel, John D. 2	''
Skinner, James 24	''
Sweet, Maryette 10	''
Sloan, Almira 17	''
Spaulding, T. E. 26	Eaton Rapids
Springer, Geo. W. 26	''
Shively, Alice 1	Lansing
Seibley, Herman 5	Millett
Stiefel, L. 9	''
Smith, Horace 21	Kingsland
Smalley, James 31	Potterville
Siberley, John 30	''
Smith, Mrs. P. 17	West Windsor
Smith, H. C. 17	''
Spears, Frank 19	''
Sabin, Mrs. M. 17	''
Sloan, James 6	''
Shotwell, L. L. 18	''
Shotwell, Albert 10	''
Terryll, D. R. 34	Dimondale
Tiedgen, A. F. 24	''
Tietgen, Dietrich 22	''
Twitchell, F. M. 13	''
Tracey, E. D. 11	''
Tracey, Chas. F. 11	''
Troupe, John 23	''
Twitchell, S. M.	''
Torrey, Chas. M. 14	''
Twitchell, Edward 24	''
Twitchell, Chas. 24	''
Traver, Chas. 1	''
Treveller, Melvin 1	''
Tuttle, Mrs. Mary 17	West Windsor
Trost, D. H. 5 & 8	''
Troub, Jerry 35	Eaton Rapids
Towsley, Geo. C. 29	Potterville
Terrill, H. F. 32	''
Terrill, Mrs. C. 29	''
Underhill, Elias	''
Underhill, Geo.	''
Urie, James 16	''
Van Osdall, John 35	''
Van Osdall, Mrs. S. 27	''
Vandervierk, S. A. 36	''
Venton, Geo. 8	''
Venton, Frank 9	''
Van Auken, C. F. 9	West Windsor
Van Auken, H. B. 21 & 30	''
Van Auken, W. H. 6	Potterville
White, Mrs. Malinda 27	Dimondale
Wright, Mrs. V.	''
Whaley, Eugene 16	''
Wakeman, S. W. 13	''
Williams, V. 21	''
Williams, Mon 1	''
Williams, Hiram 7	''
Wright, Peter 14	''
Wright, C. E. 11	''
Watrous, Chas. 11	''
Wyborn, William 25	''
Wright, Jackson 22 & 24	''
White, Samuel 27	''
Wright, Asa 35	''
Wilbur, Albert 35	Eaton Rapids
Warren, C. H. 29	West Windsor
Williams, Mrs. V. 11	Dimondale
Walford, E. N.	''
Wither, Mrs. Geo.	Potterville
Whitney, Mrs. A. 17	West Windsor
Warren, L. N. 17	''
Zacharia, Mrs.	Dimondale
Zimmerman, H. 4	''
Zimmerman, A. Estate 4	''
HAMLIN TOWNSHIP.	
Arnold, Susan M. 26	Eaton Rapids
Allyn, Austin 17 & 19	''
Allyn, Geo. 17	''
Allen, E. H. 4	''
Armstrong, Chas. 14	''
Abbey, H. F. 16	''
Brewer, L. H. 9	''
Bradford, D. J. 16	''
Bastman, William 16	''
Bledgett, Rich 16	''
Bledgett, C. L. 15	''
Bledgett, K. C. 21	''
Bledgett, I. T. 21	''
Briggs, J. H. 21	''
Beam, Mrs. M. 19	Charlesworth
Beam, John, Jr. 19	''
Bahary, Henry 18	Eaton Rapids
Bahary, Jacob 17 & 19	Eaton Rapids
Harbaugh, F. T. 15	''
Burlison, Emerson 24	''
Barrow, Henry 25	''
Beemore, Samuel 1	''
Hatterfield, Mrs. Frank M. 7	''
Bradford, Elijah 17	''
Bly, William H. 7	''
Braman, D. M. 7	''
Benton, Henry 7	''
Bidwell, Geo. 5	''
Beatty, Rolland 1	''
Bradford, Henry 18	''
Bradford, Martin 29	''
Bennett, F. C. 27	''
Burgess, Den 21	Springport
Bastman, Daniel 21	Eaton Rapids
Bastman, William F. 21	''
Beasso, Mrs. 29	Charlesworth
Betz, Lewis 23	Eaton Rapids
Bly, Mrs. Samuel 20	''
Boolster Bros. 23	''
Bishop, H. L. 11	''
Baern, S. S. 10	''
Burlison, Ellis 26	''
Clegg, Albert 34	Otter Creek
Champlan, Mrs. L. 21	Eaton Rapids
Clay, Simeon A. 21	''
Chadwick, Willis 14	''
Chadwick, Geo. 23	''
Conklin, Allen 18-21	''
Clark, Thomas 28	''
Carter, N. N. 17, 19 & 20	''
Crouse, Mrs. A.	Charlesworth
Cowan, Thomas 14 & 23	Eaton Rapids
Clay, Fred 25	''
Conklin, Jay 24	''
Cockroft, William 22 & 33	Springport
Cockroft, Levi 22 & 33	''
Collins, Jonathan 14	Eaton Rapids
Canfield, N. N. 7	''
Crittenden, S. R. 17	''
Carry, A. F. 10	''
Delamater, Lewis 11	''
Downs, Lester 19	''
Dewage, Mrs. Minerva 29	Charlesworth
Dibble, Henry 6	Eaton Rapids
De Witt, Gordon 14 & 23	''
De Witt, Corwin 23	''
De Witt, Chas. 23	''
Darling, L. B. 11	''
Evans, C. L. 31	Springport
Elmer, Orrin 5	Eaton Rapids
Egan, Henry 19	''
Freeman, Stanley 8	''
Freeman, Stillman Estate 5	''
Fuller, John T. 6	''
Fuller, Elmer 36	Otter Creek
Fuller, Truman 25	Onondaga
Fuller, Stephen 25	''
Fitzsimmons, E. L. 15	''
Fineman, Geo. 34	Eaton Rapids
Fleming, Mrs. Andrew 23	''
Fernalin, Jo 29	Charlesworth
Finch, Samuel 32	Springport
Fox, Deborah 8	Eaton Rapids
Fowler, Mrs. Emma 32	Springport
Griffith, Geo. T. 22 & 25	''
Grant, Hiram 9	Eaton Rapids
Gillman, Jacob 7	''
Gilman, I. N. 6	''
Gillman, Mrs. I. N. 7, 27 & 21	''
Gale, Fernando 5	''
Gould, D. W. 4	''
Griffith, Julius C. 7	''
Grow, Mrs. Elizabeth 10	''
Gluba, Mrs. Wallace 27	''
Glute, Chas. 23	''
Griffin, Geo. C. 31	Springport
Gifford, William H.	Eaton Rapids
Gifford, Fred 20	''
Goodyear, Luther 23	Charlesworth
Holmes, Mrs. Ann 22	Eaton Rapids
Holmes, Geo. 22	''
Holmes, Asell E. 16	''
Holmes, J. J. 10	''
Holmes, William F. 10	''
Hyatt, Chas. 8	''
Height, Mrs. Lucy 14	''
Hoag, Mrs. Anson 17-18	''
Hayward, Joel 5	''
Hart, Hiram 6	''
Hunted, Henry 5	''
Holmes, Polly 29	Charlesworth
Hershy, Silas 29	Eaton Rapids
Hershy, Nathaniel 30	''
Hammond, Ben 23	''
Hester, G. W. 21	''
Hall, Orin 16	''
Hall, Ben 21	''
Hobart, Mrs. Anna 9	''
Hamlin, D. B. 9	''
Hicks, Samuel 9	''
Hale, D. B. 10	''
Hanlon, N. P. 22	''
Howell, Nelson 3	''
Jeffries, A. J. 15	''
Knapp, Herman 24	''
Keeler, Ira 31	Onondaga
Kaylor, Jefferson 17	Eaton Rapids
Kikendall, Chas. 19 & 30	Charlesworth
Kikendall, Richard 30	''
Kikendall, James 19 & 30	''
Kikendall, Basile 21	Charlesworth
Kikendall, Geo. 19, 20 & 30	''
Kelley, Geo. T. 18	Eaton Rapids
Knight, Mrs. Geo. 9	''
Knight, Amos 9	''
Long, O. A. 9	''
Lalr, Albert 27	''
Lawrence, Ira M. 21 & 26	''
Leighton, F. S. 27, 21 & 28	''
Lindsey, Israel 19	''
London, E. A. 29	Springport
Lake, Mrs. Hulda 14	Eaton Rapids
Lake, O. B. 14	''
Lewis, John 11	''
Mitchell, Miranda 5	''
Merritt, C. E. 1	''
Markle, E. L. 26	''
Marble, W. G. 26	''
Mosley, Isaac 34	''
McManus, Geo. 24	''
McAllister, John 20	''
McLeod, William 19	Charlesworth
McAllister, William 14	Eaton Rapids
Mansfield, Christy 2	''
Montgomery, Mrs. Celia 12	''
Montgomery, Mrs. J. S. 1	''
McGilliver, Frank 24	''
Mellon, Chas. 8	''
Moon, Andrew 8	''
Mitchell, Frank 9	''
Mellon, Nancy 19	''
Miller, John 5	''
Mathias, Erias 20 & 21	Charlesworth
Nichol, George 32	''
Mendel, Frank 21	Eaton Rapids
Mathias, David 21	''
Miller, Celestia 1	''
Miller, Samuel 2	''
Noble, Geo. B. 22	''
Norris, Sophrona 13	''
Newcomb, Geo. W. 8 & 17	''
Noris, L. P. 21	''
Newcomb, Mrs. Geo. 19	Charlesworth
Noble, J. D. 16	Eaton Rapids
Neff, John 21	''
Nisbit, Archibald 17	''
Olmstead, H. D. 14	''
Olmstead, D. D. 14	''
Osborn, F. A. 15	''
Oyer, Hattie 20	Charlesworth
Parker, F. H. 4	Eaton Rapids
Pierce, Henry 3	''
Porter, Albert 4	''
Poncin, James 3	''
Pratt, William E. 7	''
Perrine, Sue 10	''
Price, James 20	''
Phillips, Geo. N. 11	''
Perrine, S. C. Estate 1 & 12	''
Perrine, D. W. 3	''
Prescott, Joseph 6	''
Pierce, D. B. 27 & 22	''
Reynolds, William W. 34	''
Rogers, Albert 27	''
Randall, Alonzo 22	''
Randall, J. H. 13	''
Rice, Almeron 22	''
Rice, Mrs. Chas. 22	''
Rice, Anna 21	''
Rhoades, Marshall 21	''
Rowe, James 21	''
Robb, Isaac 12	''
Rhoades, Robert 21	''
Ray, Joseph 4	''
Rorstock, S. H. 5	''
Rowley, Ellen M. 10	''
Rice, Almeron 21	''
Rathbun, Bruce 14	''
Reynolds, S. H. 31	''
Root, Chas. 2	''
Rice, F. W. 2	''
Ramsey, H. W. 21	''
Rorey, Jacob 18	''
Rushton, James 31	''
Rochester, Joseph 19	''
Rockwood, Robert 33	Otter Creek
Reynolds, Geo. W. 7	Eaton Rapids
Rexer, James 30	Charlesworth
Stocking, Julius 23	Eaton Rapids
Swift, Mrs. Mary 23	Otter Creek
Swift, Reuben 36	''
Swift, J. Carrington 35	''
Searls, Walsten 6	Eaton Rapids
Spear, Scott 7	''
Santo, William K. 7	''
Smith, Samuel 8	''
Swums, Eli 16	''
Seagreaves, James 1	''
Smith, B. M. 12	''
Smith, William 3	''
Schlappi, Samuel 8 & 16	''
St. John, Elias 14	''
Smith, Mrs. Elizabeth 13	''
Smith, Milo B. 13	''
Sayer, John 20 & 21	''
Smith, Chas. L. 21	''
Stump, Barbara 19 & 20	''
Shink, Mrs. Jacob 31	''
Bowles, Isaac 29	Charlesworth
Smith, William 29	''
Sayer, Geo. W. 20	''
Smith, Mrs. Emma 20	''
Sactye, Henry 31	Eaton Rapids

Spencer, Ed. 16	Eaton Rapids	Upham, Caroline 27 & 28	Eaton Rapids
Teeter, Edgar 11	"	Van Auken, Jacob 19	Charlesworth
Teeter, Mrs. C. 21	"	Van Auken, Albert 19	"
Taylor, N. T. 11	"	Van Dusen, Amos 5	Eaton Rapids
Taylor, F. D. 11	"	Vaughan, J. J. 11	"
Tilford, Robert 26	"	Vaughan, Geo. 11	"
Thoma, William 16	"	Van Allen, D. D.	"
Underdonk, W. H. 27 & 34	"	Van Gordon, Wm. 25 & 26	Otter Creek
Underwood, Levi 27	"	Walter, David 16	"

Walter, Samuel 16	Otter Creek	Wellman, R. G. 29	Eaton Rapids
Walter, Eli 15	"	Wood, Thomas 21	Springport
Walter, Chas. 10	"	Wells, J. 31	"
Walter, Marcus 1	"	Wise, Andrew 5	Eaton Rapids
Wisner, C. D. 16	"	Wood, David 5	"
Wheeler, Mrs. Lucretia 21	"	Wood, Job 6	"
Wheeler, L. J. 22	"	Willoby, Samuel 5	"
Wheeler, Mrs. Orwa 22	"	Weldon, C. H. 5	"
Wyman, Milton B. 21	"	Williams, Mrs. L. 28	"
Williams, Chas. 13	Charlesworth	Williams, Owen 26	"

Wolcott, M. S. 24	Eaton Rapids
Wright, V. N. 13	"
Wright, Walter 16	"
Williams, James 17	"
Wise, Mrs. Melvina 17	"
Whitaker, Geo. 19	Charlesworth
Wilbur, H. 33	Springport
Wilbur, F. S. 31	"
Wilson, W. W. 31 & 35	"

ELECTIONS.

1840—Harrison, Whig, 837; VanBuren, Democrat, 775. Total, [illegible]. 1844—Clay, Whig, 410; Polk, Democrat, 376; Birney, Liberty, 5. Total, 791. 1848—Taylor, Whig, 316; Cass, Democrat, 146; VanBuren, Freesoil, 718. Total, 1190. 1852—Scott, Whig, 637; Pierce, Democrat, 796; Hale, Freesoil, 275. Total, 1618.

1856.

	Fremont.	Buchanan.	Fillmore
Bellevue	151	110	8
Benton	94	43	0
Brookfield	65	35	9
Carmel	142	126	8
Chester	81	47	0
Delta	94	31	1
Eaton	118	140	0
Eaton Rapids / Hamlin	228	265	8
Kalamo	98	126	0
Oneida	187	45	4
Roxand	133	36	0
Sunfield	51	18	0
Vermontville	130	29	0
Walton	133	36	0
Windsor	91	33	0

1860.

	Lincoln.	Douglas.	Bell.
Bellevue	204	160	0
Benton	96	66	0
Brookfield	99	48	0
Carmel	195	131	0
Chester	100	69	0
Delta	76	43	0
Eaton	187	153	0
Eaton Rapids / Hamlin	337	238	0
Kalamo	114	106	0
Oneida	235	77	0
Roxand	109	38	0
Sunfield	73	15	0
Vermontville	146	40	0
Walton	171	55	1
Windsor	95	83	0

1864.

	Lincoln.	McClellan.
Bellevue	183	171
Benton	84	67
Brookfield	84	86
Carmel	135	147
Chester	74	63
Delta	98	54
Eaton	198	149
Eaton Rapids / Hamlin	261	266
Kalamo	119	95
Oneida	199	87
Roxand	109	81
Sunfield	98	97
Vermontville	183	43
Walton	154	88
Windsor	84	78

1868.

	Grant.	Seymour.
Bellevue	290	229
Benton	144	108
Brookfield	120	77
Carmel	305	127
Chester	128	97
Delta	118	95
Eaton	247	204
Eaton Rapids / Hamlin	480	366
Kalamo	173	129
Oneida	277	104
Roxand	166	50
Sunfield	113	39
Vermontville	227	48
Walton	250	64
Windsor	150	108

1872.

	Grant.	Greeley.	O'Connor.	Black.
Bellevue	318	174	5	25
Benton	194	78	0	10
Brookfield	114	48	0	4
Carmel	115	91	0	5
Chester	148	83	2	8
Delta	187	40	0	18
Eaton	109	97	0	0
Eaton Rapids	341	207	16	17
Hamlin	214	109	0	7
Kalamo	190	98	4	11
Oneida	256	155	1	40
Roxand	130	89	7	1
Sunfield	125	62	1	0
Vermontville	243	96	1	4
Walton	295	55	4	12
Windsor	150	73	14	4
Charlotte	464	116	3	46

1876.

	Hayes.	Tilden.	Cooper.	Smith.
Bellevue	287	243	0	9
Benton	219	110	0	4
Brookfield	184	188	22	0
Carmel	189	159	3	5
Chester	165	161	11	8
Delta	204	123	11	0
Eaton	164	185	0	2
Eaton Rapids	287	216	0	62
Hamlin	261	174	0	11
Kalamo	238	182	1	10
Oneida	401	292	18	1
Roxand	247	97	0	0
Sunfield	189	99	0	0
Vermontville	224	143	0	8
Walton	298	135	0	9
Windsor	240	155	4	1
Charlotte	417	285	1	80

1880.

	Garfield.	Hancock.	Weaver.	Dow.
Bellevue	329	210	47	9
Benton	273	165	22	0
Brookfield	155	109	56	8
Carmel	180	149	16	3
Chester	191	142	7	3
Delta	194	86	110	4
Eaton	163	145	8	11
Eaton Rapids	297	290	13	60
Hamlin	254	175	10	95
Kalamo	220	179	17	10
Oneida	246	125	274	4
Roxand	251	58	63	49
Sunfield	203	124	14	0
Vermontville	291	156	15	0
Walton	328	150	5	6
Windsor	282	111	73	6
Charlotte	482	235	58	18

1884.

	Blaine.	[Cleveland.	Butler.]	St. John.
Bellevue	290	236		16
Benton	201	140		19
Brookfield	182	183		18
Carmel	176	169		11
Chester	170	156		19
Delta	150	183		23
Eaton	156	170		14
Eaton Rapids	127	133		23
Hamlin	163	108		27
Kalamo	233	179		9
Oneida	283	490		33
Roxand	230	143		45
Sunfield	216	139		18
Vermontville	265	224		22
Walton	300	161		51
Windsor	213	235		13
Charlotte	490	287		21
Eaton Rapids	254	294		60

1888.

	Harrison.	Cleveland.	Fisk.	Streeter.
Bellevue	308	254	4	3
Benton	241	155	63	36
Brookfield	183	164	4	14
Carmel	162	163	1	17
Chester	191	164	6	16
Delta	208	183	30	40
Eaton	171	187	0	15
Eaton Rapids	156	172	8	22
Hamlin	122	99	1	45
Kalamo	241	190	0	15
Oneida	232	224	217	28
Roxand	245	171	38	83
Sunfield	247	105	34	22
Vermontville	201	208	0	36
Walton	320	164	0	44
Windsor	217	193	21	17
Charlotte	566	305	1	170
Eaton Rapids	304	221	4	60

1892.

	Harrison.	Cleveland.	Weaver.	Bidwell.
Bellevue	254	241	15	11
Benton	205	122	89	22
Brookfield	109	144	40	10
Carmel	172	118	8	11
Chester	168	129	6	6
Delta	170	119	88	20
Eaton	123	118	8	18
Eaton Rapids	112	101	47	14
Hamlin	194	76	87	40
Kalamo	181	108	15	28
Oneida	279	277	[illegible]	86
Roxand	197	98	20	18
Sunfield	211	113	65	25
Vermontville	274	178	6	44
Walton	221	191	30	41
Windsor	179	107	62	20
Charlotte	506	409	55	55
Eaton Rapids	271	225	19	43

www.ingramcontent.com/pod-product-compliance
Lightning Source LLC
Chambersburg PA
CBHW021943190326
41519CB00009B/1119
* 9 7 8 3 7 4 3 4 2 3 9 3 0 *